Neurotechnology and Direct Brain Communication

Neurotechnology and Direct Brain Communication focuses on recent neuroscientific investigations of infant brains and of patients with disorders of consciousness (DOC), both of which are at the forefront of contemporary neuroscience. The prospective use of neurotechnology to access mental states in these subjects, including neuroimaging, brain simulation, and brain computer interfaces, offers new opportunities for clinicians and researchers, but has also received specific attention from philosophical, scientific, ethical, and legal points of view. This book offers the first systematic assessment of these issues, investigating the tools neurotechnology offers to care for verbally non-communicative subjects and suggesting a multidisciplinary approach to the ethical and legal implications of ordinary and experimental practices.

The book is divided into three parts: the first and second focus on the scientific and clinical implications of neurological tools for DOC patient and infant care. With reference to these developments, the third and final part presents the case for re-evaluating classical ethical and legal concepts, such as authority, informed consent, and privacy.

Neurotechnology and Direct Brain Communication will appeal to researchers and postgraduate students in the fields of cognitive science, medical ethics, medical technology, and the philosophy of the mind. With implications for patient care, it will also be a useful resource for clinicians, medical centres, and health practitioners.

Michele Farisco is part of the Neuroethics team of the Centre for Research Ethics and Bioethics, Uppsala University, Sweden, where he is completing his second PhD; member of the Ethics & Society Program in the European Human Brain Project, where he is involved in the research in philosophy and neuroethics; and head of the Science and Society Research Unit, Biogem Genetic Research Centre, Ariano Irpino, Italy. He is the author of three books and several articles in the areas of posthuman philosophy and philosophical, ethical, and legal implications of genetics and neuroscience.

Kathinka Evers is Professor of Philosophy and Senior Researcher at the Centre for Research Ethics and Bioethics, Uppsala University, Sweden, and Honorary Professor at the Central University of Chile, Chile. She is co-director of the Ethics & Society Program in the European Human Brain Project, leading the research in philosophy and neuroethics. Formerly the Executive Director for the Standing Committee on Responsibility and Ethics in Science (SCRES) of the International Council for Science (ICSU), her main research interests are in neuroethics and neurophilosophy, with special focus on analyses of consciousness and brain simulation.

Explorations in Cognitive Psychology Series

Books in this series:

Perception Beyond Gestalt
Progress in vision research
Edited by Adam Geremek, Mark Greenlee, and Svein Magnussen

Fine Art and Perceptual Neuroscience
Field of vision and the painted grid
Paul M.W. Hackett

Simulation Theory
A psychological and philosophical approach
Tim L. Short

Neuropsycholinguistic Perspectives on Language Cognition
Essays in honour of Jean-Luc Nespoulous
Edited by Corine Artésano and Mélanie Jucla

Contextualizing Human Memory
An interdisciplinary approach to understanding how individuals and groups remember the past
Edited by Charles B. Stone and Lucas Bietti

Neurotechnology and Direct Brain Communication
New insights and responsibilities concerning speechless but communicative subjects
Edited by Michele Farisco and Kathinka Evers

Neurotechnology and Direct Brain Communication

New insights and responsibilities concerning speechless but communicative subjects

Edited by Michele Farisco and Kathinka Evers

Routledge
Taylor & Francis Group

LONDON AND NEW YORK

First published 2016 by Routledge

2 Park Square, Milton Park, Abingdon, Oxfordshire OX14 4RN
711 Third Avenue, New York, NY 10017

Routledge is an imprint of the Taylor & Francis Group, an informa business

First issued in paperback 2017

British Library Cataloguing in Publication Data
A catalogue record for this book is available from the British Library

Library of Congress Cataloging in Publication Data
Names: Farisco, Michele, author. | Evers, Kathinka, author.
Title: Neurotechnology and direct brain communication : new insights and responsibilities
concerning speechless but communicative subjects / Michele Farisco, Kathinka Evers.
Description: Abingdon, Oxon ; New York, NY : Routledge, 2016.
Identifiers: LCCN 2015046135| ISBN 9781138851672 (hardcover) |
ISBN 9781315723983 (ebook)
Subjects: LCSH: Facilitated communication. | Neurotechnology (Bioengineering) |
People with disabilities—Means of communication. | Consciousness.
Classification: LCC RC429 .F37 2016 | DDC 616.855—dc23
LC record available at http://lccn.loc.gov/2015046135

ISBN: 978-1-138-85167-2 (hbk)
ISBN: 978-0-8153-6062-9 (pbk)

Typeset in Galliard
by diacriTech, Chennai

To my beloved daughter Ilde: you're teaching me the beauty of the mind's sunrise.

Michele Farisco

Contents

Acknowledgments   ix
List of contributors   xi

Introduction: exploring a speechless world   1
MICHELE FARISCO AND KATHINKA EVERS

PART I

1 **The emergence of consciousness: from foetal to newborn life**   7
   HUGO LAGERCRANTZ AND NELLY PADILLA

2 **Mapping mind-brain development**   21
   ANDREAS DEMETRIOU, GEORGE SPANOUDIS, AND MICHAEL SHAYER

3 **Cognitive capacities of the infant mind: a neuroimaging perspective**   40
   MOHINISH SHUKLA AND VIVIAN CIARAMITARO

4 **Neural infantese: detecting pain and suffering in preverbal infants by means of neuro-technological communication**   51
   KARL SALLIN

PART II

5 **Instrumental assessment of residual consciousness in DOCs**   69
   CARLO CAVALIERE, CAROL DI PERRI, STEVEN LAUREYS, AND ANDREA SODDU

6 Neurotechnological communication with patients with disorders of consciousness 85

DAMIEN LESENFANTS, CAMILLE CHATELLE, JAD SAAB, STEVEN LAUREYS, AND QUENTIN NOIRHOMME

7 Does task-evoked activity entail consciousness in vegetative state? "Neuronal-phenomenal inference" versus "neuronal-phenomenal dissociation" 104

GEORG NORTHOFF

PART III

8 Ethical and deontological issues in paediatric clinical studies: an analysis of documents from national and international institutions 119

CARLO PETRINI

9 Disorders of consciousness and informed consent 133

RALF J. JOX

10 Brain-imaging and privacy concerns 143

ARLEEN SALLES

Conclusion 157

MICHELE FARISCO AND KATHINKA EVERS

Index 161

Acknowledgments

Michele Farisco and Kathinka Evers are supported by funding from the European Union Seventh Framework Programme (FP7/2007–2013) under grant agreement n° 604102 (HBP).

List of contributors

Carlo Cavaliere
Coma Science Group, Cyclotron Research Center & Neurology Department, University of Liège, Belgium
IRCCS SDN, Istituto Ricerca Diagnostica Nucleare, Naples, Italy

Camille Chatelle
Coma Science Group, Cyclotron Research Centre and Neurology department, University of Liège, Liège, Belgium
Department of Physical Medicine and Rehabilitation, Spaulding Rehabilitation Hospital, Boston, Massachusetts, USA
Laboratory for NeuroImaging of Coma and Consciousness, J. Philip Kistler Stroke Research Center, Massachusetts General Hospital, Boston, Massachusetts, USA

Vivian Ciaramitaro
Department of Psychology, College of Liberal Arts, University of Massachusetts Boston, Boston, MA, USA

Andreas Demetriou
Psychology of Social Sciences, University of Nicosia, Cyprus

Carol Di Perri
Coma Science Group, Cyclotron Research Center & Neurology Department, University of Liège, Belgium

Kathinka Evers
Centre for Research Ethics and Bioethics, Uppsala University, Sweden

Michele Farisco
Centre for Research Ethics and Bioethics, Uppsala University, Sweden
Biogem Genetic Research Centre, Ariano Irpino (AV), Italy

Ralf J. Jox
Institute of Ethics, History and Theory of Medicine, Ludwig-Maximilians University of Munich, Germany

Hugo Lagercrantz
Karolinska Institute and Astrid Lindgren Children's Hospital, Karolinska University Hospital, Stockholm, Sweden

Steven Laureys
Coma Science Group, Cyclotron Research Center & Neurology Department, University of Liège, Belgium

Damien Lesenfants
Coma Science Group, Cyclotron Research Centre and Neurology Department, University of Liège, Liège, Belgium
School of Engineering and Institute for Brain Science, Brown University, Providence, Rhode Island, USA

Quentin Noirhomme
Coma Science Group, Cyclotron Research Centre and Neurology department, University of Liège, Liège, Belgium
Brain Innovation B.V., Maastricht, the Netherlands

Georg Northoff
University of Ottawa Institute of Mental Health Research, Ottawa, Canada
Center for Cognition and Brain Disorders, Hangzhou Normal University, Hangzhou, China
Center for Brain and Consciousness, Taipeh Medical University (TMU), Taipeh, Taiwan
College for Humanities and Medicine, Taipeh Medical University (TMU), Taipeh, Taiwan
Royal Ottawa Healthcare Group, University of Ottawa Institute of Mental Health Research, Ottawa, Canada

Nelly Padilla
Karolinska Institute and Astrid Lindgren Children's Hospital, Karolinska University Hospital, Stockholm, Sweden

Carlo Petrini
Bioethics Unit, Office of the President, Istituto Superiore di Sanità (Italian National Institute of Health), Rome, Italy

Jad Saab
School of Engineering and Institute for Brain Science, Brown University, Providence, Rhode Island, USA

Arleen Salles
Centre for Research Ethics & Bioethics (CRB), Uppsala University, Sweden
CIF, Center of Philosophical Investigations, Buenos Aires, Argentina

Karl Sallin
Centre for Research Ethics & Bioethics (CRB), Uppsala University, Sweden
Department of Women's and Children's Health, Division of Neonatology, Karolinska Institute, Solna, Sweden

Michael Shayer
Kings College, University of London, UK

Mohinish Shukla
Department of Psychology, College of Liberal Arts, University of Massachusetts
Boston, Boston, MA, USA

Andrea Soddu
Coma Science Group, Cyclotron Research Center & Neurology Department,
University of Liège, Belgium
Physics and Astronomy Department, Brain and Mind Institute, University of
Western Ontario, Canada

George Spanoudis
Psychology Department, University of Cyprus, Cyprus

Introduction

Exploring a speechless world

Michele Farisco and Kathinka Evers

The instrumental investigation of consciousness has witnessed an astonishing progress over the last years. Different neurotechnological tools and methods have been developed in order to assess consciousness in not yet verbally communicative subjects such as infants and to assess residual consciousness in no longer verbally communicative subjects such as patients with Disorders of Consciousness (DOCs). Neuroimaging technologies, particularly the functional neuroimaging technologies, give us the possibility to see what happens in the brain during the execution of particular tasks. Several important conceptual issues arise from neurotechnological assessments of consciousness, particularly concerning the connection between cerebral structure and architecture on the one hand and the conscious mind on the other. Different interpretations of their relationship have been suggested, oscillating between two opposite extremes, arguing either that mind and consciousness can be equated to the brain or that mind and consciousness are substantially different from the brain. From a scientific and clinical point of view, a trend toward the former extreme, even if differently modulated, has been essential in order to reach a better understanding of consciousness, a better understanding of how consciousness emerges, and a better management of people suffering from DOCs.

Notably, the identification of the activated areas and the real-time observation of the cerebral activity potentially allow a new form of technology-based communication without first-person overt behavior or speech, thus going beyond the behavioral manifestation of awareness. This technology-mediated investigation of the inner-world of non-behavioral or non-verbal subjects potentially opens new, unexplored fields of discussion, involving science, philosophy, ethics, and law.

Beside the technical feasibility of this new kind of communication, it is important to explore and assess the conceptual and ethical issues emerging from it. First of all, what does it mean to communicate in this context? The scientific exploration of the brain, assumed as the biological essential underpinning of language and communication, is giving us new data and new paradigms for exploring what is behind our communicative abilities and practices. The relevance of this scientific investigation is clear: without a minimally functioning brain, we could not communicate, or even formulate any message. It is therefore important to study the brain and explore its networks and functions involved in communication, especially, but not only, for therapeutic purposes.

What is at stake is the explanatory ability of neuroscience: to what extent, and in what way can we explain communication via brain architecture? Can we develop a bridge between the cerebral structure and the semantic complexity of language? In other words, is it possible to read or interpret what an individual thinks or feels directly from the activation of the different brain regions? Conceptually these issues bring us back to what in philosophy of mind is defined as the "hard problem," i.e., the relationship between brain and mind: passionate and hard discussions have been dedicated to this problem, which is maybe unsolvable.

In this book we do not tackle this general conceptual problem directly. We prefer to start from the most advanced stage of the scientific investigation of cerebral structure and function and from the most sophisticated technological applications and then reflect about their possible impacts. First of all we suggest that neuroscience and neurotechnology are conceptually relevant in two basic senses: they are not neutral from a conceptual point of view, because they start from particular premises which deserve specific attention; and they can affect the traditional way to think about notions like mind, language, and consciousness. For this reason it is important to critically assess the neuroscientific investigation of human mind and to rethink some traditional notions in the light of its results.

Furthermore, this book is dedicated not only to the aforementioned conceptual issues, but also to the ethical and legal implications of neuroscience and neurotechnology. Among other possible emerging issues, informed consent, privacy, and integrity are some of the most challenging and problematic, and they deserve specific attention and investigation, as exemplified in the last part of the book.

The specific focus of the following chapters is the brain of infants and of patients with DOCs. We strategically choose these extreme stages, the edges of the human life, in order to scrutinize the possibility that the sunrise and the sunset of mind could give us a key to solve the enigma of the ordinary mind. Moreover, both infant and subjects with DOCs are unable to verbally communicate but anyway retain a living, active brain, that is still not mature in the first case or deeply damaged in the second case, but in both cases able to sustain an activity potentially resulting in interaction with the external environment, i.e., in communication.

Neuroscience and neurotechnology promise to give new, innovative tools for exploring these "edge" conditions, with possibilities to improve their care and management. This book specifically focuses on new possibilities to communicate with these speechless subjects through the mediation of technology. In the first case (i.e., with infants) it is perhaps questionable to talk about a proper communication: we could imagine to use technology to directly check in their brains the reaction to particular external or internal conditions (i.e., environmental settings or the use of specific drugs) and then infer their needs, but since they lack the ability to understand language, it is more problematic to think of them as able to intentionally respond to verbal requests from the external world. Nevertheless, they have active brains, well able to react to external stimuli and to express needs and preferences resulting from their internal states: the point is how to technologically detect this speechless expression, prospectively improving the quality of the first months of life, especially, but not only, in the case of infants with particular diseases and disabilities. The first part of the present book specifically addresses these points, showing how important progress has been made in the particular study of the infant brain and mind, as well as in the general understanding of the human brain and mind starting from infant subjects.

Patients with DOCs whose brains are very damaged have been able to meaningfully react to external stimulation in some experimental conditions. In this case it is reasonable to think about a minimal or rudimental form of communication, which, if confirmed, is very relevant not only from a scientific and technical, but also from an ethical and legal point of view. The second part of the book offers a deep investigation of what is technically possible today for investigating residual consciousness in patients with DOCs and what this implies for a prospective communication with them.

So, why focus on infants and subjects with DOCs? There are at least three reasons for that.

First of all, infants and brains of patients with DOCs are extremely interesting in themselves. For instance, brain development processes after birth are still to be cleared in details both regarding biological alterations and in relation to evolving behavior. Moreover, it is scientifically evident that infant brain has much more synapses than the adult brain, even up to twice as many synapses in average: does this imply that they are differently conscious? A possibility is to conceive infants in a kind of third state between being fully aware and fully unaware, a particular condition that we have lost in which the sense of self is mixed up with the awareness of other people. Analogously, the residual consciousness of patients with DOCs could be seen as a different level of healthy consciousness or as a different kind of consciousness. Does this distinction (between level and kind of consciousness) make sense? If a disordered consciousness is so deeply different than the healthy one (both in terms of level and/or in terms of kind), how could we communicate with patients with DOCs?

Second, infant and brains of patients with DOCs are interesting in connection to each other. In both cases we probably have a partial self-awareness and a less analytical mind than the adult healthy human condition: this means that a similar "opaque" mental condition is shared by infants and patients with DOCs, so that cues about one could be helpful to understand the other. Moreover, similar unconscious priming and "blindsight" can occur both in infants and (impaired) adults. Another important potential similarity between infants and patients with DOCs is that in both cases the absence of responsiveness cannot be assumed as evidence of unawareness: we could have the same condition of not responding aware state. Moreover, in both cases consciousness cannot be reduced to cognition, i.e., to cortical areas activation: it is necessary to stress the importance of the emotional dimension of consciousness as well, i.e., to focus on subcortical cerebral areas. The relevant point for the present book is: how to communicate with this "primordial" emotional consciousness?

Finally, to study the brain of infants and of patients with DOCs is potentially useful for better understanding brain and consciousness as such. These edge conditions can potentially give us important cues in our search for the so-called signatures of consciousness.

As showed in the following, the neural evoked activity in patients with DOCs is not an unproblematic sign of consciousness, at least for two reasons: the inference from neurons to phenomenal experience is problematic; the role of resting state needs to be taken into account. The same is true for healthy people. Yet the brain of patients with DOCs could give new insights about the connection between neural markers and subjective experience.

Concerning infants, the cognitive continuity between them and adults is increasingly revealed, so to study the first is surely useful for understanding the second. Particularly the less mature and more plastic brain of infants deserves specific attention, not only because it is interesting in itself, but also because it could give us new conceptual tools to understand the plastic brain in general.

Part I

1 The emergence of consciousness
From foetal to newborn life

Hugo Lagercrantz and Nelly Padilla

What is it like to be a baby?

What is it like to be a bat? This important question was evoked by Thomas Nagel in 1974 and has then been discussed by philosophers over and over again (Nagel, 1974). Nagel chose the bat as an example since it is so different from the human. It hangs upside-down and uses sonar to communicate, etc. But what is it like to be a human foetus or a baby? The foetus is also usually positioned upside-down and listens to the filtered voice of the mother. The newborn infant spends most of its time in horizontal position and starts to imitate the facial expressions of adults and absorbs phonemes, which actually begins before birth. To what extent the newborn infant is conscious and how new neurotechnologies can be used to measure infant consciousness is discussed in this updated version of previous reviews (Lagercrantz and Changeux, 2009; Lagercrantz, 2014). The emergence of consciousness is related to the neurobiological and psychological development of the brain (Lagercrantz et al., 2010). This question, which may be elusive, does have important clinical implications to predict neuropsychiatric disease like autistic spectrum disorder (ASD) using neuroimaging biomarkers early in life to identify infants at risk and provide targeted intervention at an early stage.

Definition of consciousness

According to Henri Bergson, the primary function of consciousness is to retain what no longer is and to anticipate what as yet is not (see Posner and Rothbart, 1998). This definition of consciousness is not applicable to the newborn baby. However, the baby seems to be conscious if this is defined as awareness of the body, oneself, and the outside world, at least at a minimal level (Zelazo, 2004). Thus the infant seems to be conscious of something and can react with avoidance or cry. In this regard, using functional magnetic resonance imaging (fMRI) it has been demonstrated that the newborn has activated brain networks oriented toward sensory systems (Fransson et al., 2011), which contribute to the sense of body ownership.

It is important to distinguish between the states of consciousness, i.e., wakefulness, sleep, coma, and general anaesthesia, versus the content of consciousness. There is some controversy concerning whether rapid-eye movement (REM) sleep with dreaming should be regarded as a conscious or non-conscious state. Since purposeful movements are usually not performed and cortex is not activated to the same extent as during wakefulness, it should be regarded as an essentially unconscious state. Furthermore, insight and self-reflection are absent during dreams.

Models

There are several theoretical models of consciousness. The Integrated Information Theory (IIT) has been proposed by Giulio Tononi (Laureys and Tononi, 2008). It postulates that one can be conscious of multiple things and that they are highly integrated. A number of neuronal circuits are involved in the integration of all the conscious experiences. This can be further estimated mathematically.

Another model has been proposed by Jean-Pierre Changeux and Stanislas Dehaene (Changeux and Dehaene, 1989; Dehaene, 2014). Whenever we become conscious about something it can be retained in the working memory. It can then be processed in the global neuronal workspace (GNW), a number of long neurons interconnecting various hubs in the brain. In this way the impression from any sense organ, such as a familiar face or voice, a taste or a smell, can be associated with old memories and integrated.

A third model of consciousness is proposed from the neuroanthropology approach (Northoff, 2010), which emphasizes the huge importance of cultural circuits in the formation of individual consciousness (Bartra, 2014). Brain circuits may use symbolic resources from the cultural networks in their different conscious operations constituting bidirectional traffic between subjects and environment that is reflected in the brain's neuronal activity (Han and Northoff, 2008). In infants, for example, the development of spoken language begins as a social activity, and then it acquires a self-centred character to finally generate inner speech. Instead, in autism, the dysfunctional connection of the brain with the sociocultural circuits results in difficulties establishing social relations and verbal and non-verbal communication. Hence, structural and functional brain development will require extending networks from the brain through the body into the sensorimotor environment (Byrge et al., 2014). Thus, in order to reach a high-level consciousness (or self-consciousness) the individual and their internal sensations must be exposed to the external world (Bartra, 2014).

Neurotechnological assessment

The hard problem (Chalmers, 1996) is whether it is possible to bridge the hump from what is going on at a neurophysiological level to subjective feeling. This is particularly difficult in infants who cannot report what they are conscious about. By magnetic resonance imaging (fMRI) the difference between oxygenated and deoxygenated blood can be monitored. This blood oxygen level-dependent signal (BOLD) is assumed to correlate with neuronal activity, particularly the slow cortical potentials (SCP) (He and Raichle, 2009), which has been proposed to correspond to "the stream of consciousness". However, there is an inverse-signal response in infants probably due to a lower increase of perfusion as compared with oxygen consumption (Born et al., 1996).

The problem with fMRI particularly when monitoring infants is that the head has to be immobilized. Furthermore, there is considerable noise. To overcome these difficulties fast acquisitions and advanced MRI sequences have been developed. fMRI is used to study normal and abnormal patterns of brain activations and also to evaluate the activity of the brain in its resting state. This methodology has been successfully applied to examine neonatal populations (Doria et al., 2010) defining patterns of neural networks development in the maturing brain. Recently, the application of an advanced processing method to evaluate fMRI (dynamic functional connectivity) has demonstrated that wakefulness is characterized by a great diversity of brain states, which may constitute a signature of consciousness (Barttfeld et al., 2015). This promising methodology may provide unique insight into early functional brain development related to consciousness in infants.

A simpler method to assess the hemodynamic response is to use Near Infrared spectroscopy (NIRS). The NIRS device produces near infrared light at different wavelengths. This is reflected by natural chromophores like oxygenated and deoxygenated haemoglobin. The signals are picked up by receiver probes and correlate to the blood flow which is assumed to correspond to the neuronal activity in areas of interest. By this way the somatosensory response to painful procedures has been recorded in preterm infants as an indication of awareness of pain (Bartocci et al., 2006). NIRS is less sophisticated than fMRI, but can be used bedside and is a fairly silent method.

Electroencephalography (EEG) can also be used, particularly to monitor event-related potentials (ERP) in response to sound and visual impressions. ERP was monitored in 5- to 15-month-old infants by showing them faces that were masked to render them visible or non-visible (Kouider et al., 2013). A late ignition in the ERP response was found already at 5 months assumed to reflect conscious perception. However, the response was fairly weak as compared with the older infants. ERP is also useful to study how speech and language are processed in the infant brain (Kuhl, 2010).

EEG can also be used to monitor the resting-state activity, by modelling the frequency power spectrum with a power-law function (Fransson et al., 2013). Similar patterns of connectivity in the infant brain have been observed with this method as by using fMRI.

Magnetoencephalography (MEG) measures the magnetic fields associated with electrophysiological brain activity which is a more direct method than analyzing the hemodynamic response with fMRI. It may disturb the infant less than fMRI, since it is less noisy and can be performed more rapidly. It can be used both for human foetuses and neonates (Schleger et al., 2014).

Neural correlates of consciousness

Human consciousness is assumed to be related with neural activity in the brain and particularly in the neocortex (Koch, 2004). According to Sporns (Heuvel and Sporns, 2011), "the collective actions of individual nerve cells linked by a dense web of intricate connectivity guide behaviour, shape thoughts, from and retrieve memories, and create consciousness". The "atoms" of consciousness, i.e., the neurons, prolipherate mainly between the tenth to the twentieth gestational week (see Lagercrantz et al., 2010). There does not seem to be any neurogenesis in the cortex after birth (Nowakowsky, 2006). The neurons begin to sprout, develop dendritic spines during the third trimester. Synaptogenesis is also ignited during this period to peak at about one year after birth (Bourgeiois, 2010). The "synaptic crosstalk" between the neurons is of course essential for consciousness. In this way, single neurons are interconnected into coherent population and integrated into systems that enable local regions to participate of dynamic events providing the functional bases for consciousness.

A prerequisite for the emergence of consciousness is also that the thalamocortical connections have developed (Kostovic and Judas, 2010). The neurons from the sensory organs (except olfaction) terminate in the subplate of the cortex before about 25 weeks of gestational age. The subplate may be up to four times thicker than the cortical plate and serves as a waiting zone and a guidance hub for the afferents from the thalamus and other areas of the brain. After that the ingrowth of thalamocortical axons in the somatosensory, auditory, visual, and frontal cortex commences. However, most of the corticocortical neuronal circuits develop later during infancy. In summary, the development of connections between the thalamus and the cortex as early as 25 weeks of gestation creates the structural substrate for various sensory experiences in the newborn.

The immature brain circuitries are reflected by the pattern of the EEG. This has been found to be discontinuous in preterm infants interrupted by so called spontaneous activity transients (SATs) or spindle bursts, which can be observed at about 23 to 24 weeks of gestation (Vanhatalo and Kaila, 2006). They may be generated by immature neurons establishing their connectivity. This is also reflected by the patterns of the somatosensory evoked potentials (Vanhatalo and Lauronen, 2006). With the emergence of consciousness, the characteristic activity pattern of the EEG during wakefulness is generated by the corticothalamic system. In addition, groups of cells in the brainstem and hypothalamus provide neuromodulatory input to the cortex that promotes arousal (Deco et al., 2014).

Spontaneous resting state activity

Even without any task performance there is a low frequency spontaneous intrinsic brain activity called the resting state activity. It generates patterns of functional connectivity involving different regions of the brain exhibiting a spatial organization into functional networks very well described in adults. Resting-state networks have also been found in the foetal brain (Schöpf et al., 2012), the preterm infants at term age (Fransson et al., 2007), and full-term infants soon after birth (Fransson et al., 2009). Thus, the newborn brain cannot be regarded as a blank slate as previously believed.

The topology of structural and functional networks change throughout development driving behaviour which in turn can change pattern of connectivity modulating future behaviour (Byrge et al., 2014). In this way, consciousness as the waking state, as experience, and as mind is not a static product but an evolving characteristic. An atypical organization of functional networks and its extension into the world is related with many developmental disorders (Byrge et al., 2014). For example, an atypical neural representation of the self is a key mechanism of both self-referential and social impairments in autism (Lombardo et al., 2010).

Default mode network

The default mode network, a set of regions that participate in internal modes of cognition, may subserve the mechanisms for maintaining and promoting consciousness (Demertzi et al., 2014). This network shows specific connectivity changes as the level of consciousness diminishes, thus connectivity strength could be an indicator of the level of consciousness under different conditions (Vanhaudenhuyse et al., 2010).

In infants, the default mode network shows a temporal evolution with an immature and incomplete network in neonates (Fransson et al., 2009; Doria et al., 2010), and a more complex and intensively connected network at one year of life. By two years, the default mode network shows similar structural characteristics as the adult network. Despite the different structures composing the default mode network during infancy, the posterior cingulate cortex/retrosplenial represents one of the main hubs of the default mode network as in adults (Gao et al., 2009). Indeed, the posterior cingulate cortex supports a central role in the development of functional brain networks already during foetal life when negative correlations between this structure and other parts of the brain become more negative with advancing gestational age (Thomason et al., 2014). Such negative correlations are absent in the anesthetized brain and are a genuine characteristic in the awake brain (Barttfeld et al., 2015).

It has been proposed that the structural organization of the default mode network may predate its functional specialization (Gao et al., 2009), even though the temporal and spatial evolution of the default mode network seems to be correlated with the evolving trajectory of self-consciousness in infants before two years of life.

Psychological criteria of consciousness

Sensory awareness is a crucial component of consciousness (Koch, 2004) which is also amenable to investigate even in the infant. This involves the ability to be awake and aware of visual, auditory, olfactory, and sensory factors including painful impressions. Other important components are memory and social communication.

All these criteria can be mimicked by a virtual baby (Cotterill, 2003). To achieve consciousness a subjective component is also required. Emotional feelings like joy (Kringelbach and Berridge, 2012), love, hatred, and sadness are also important components of consciousness.

Awakefulness

A common-sense definition of consciousness, according to John Searle, is those states of sentience and awareness that typically begin when we awake from a dreamless sleep and continue until we go to sleep again, or fall into coma or die or otherwise become "unconscious" (Searle, 2000). With this definition the foetus and the preterm infant born before 24 weeks are never conscious, although the foetus and newborn infant spend most of their time in active sleep and are thus potentially dreaming. However, dreaming is tightly linked to the ability to imagine things visually, which is less likely to occur in the foetus and the extremely preterm infant.

The foetus is living at a very low oxygen level ("Mount Everest in utero"), which probably suppresses foetal activity by increasing the level of adenosine. This degradation product of ATP acts as a sedatory neuromodulator. The level of the neurosteroid pregnenolone produced by the placenta is more than ten fold higher in foetal blood (Mellor et al., 2005). Prostaglandin E2 occurs also in higher levels and sedates the foetus.

On the other hand the important neurotransmitter GABA is excitatory during early foetal life (see Lagercrantz, 2010). Thus there is probably a high activity in the foetal brain, which is of importance for the neuronal wiring. This noise in the foetal brain is related to dream activity but lacks integration and coherence and is less likely to generate consciousness.

The newborn infant seems to be awakened at birth. The baby's eyes become wide open with usually large pupils and it may cry. This arousal is probably triggered by the stress of being born (Lagercrantz and Slotkin, 1986; Lagercrantz, 1996) and the transition from the warm environment in the womb to the cooler extrauterine environment. The evaporation of amniotic fluid has this cooling effect (Gluckman, 1983) even in a tropical milieu. This also triggers the first breaths of air. The first breaths of air have since the antique time been regarded as the ignition of life as indicated by the word *spiritus*.

The stress of being born is probably mainly due to the squeezing and squashing of the foetal head during vaginal delivery. This triggers an enormous catecholamine surge resulting in about a 20-fold higher level in umbilical arterial blood (Lagercrantz and Slotkin, 1986). There is probably a parallel surge of the noradrenergic activity in the brain originating from the locus coeruleus, as indicated by studies of newborn rats (Lagercrantz et al., 1992).

Locus coeruleus is responsible for the arousal. After elective caesarean section less catecholamines are released, which also has been showed to delay the transition at birth. However, the cooling and also the clamping of the umbilical cord and the removal of the mentioned placental suppressors seem to be sufficient to awake the newborn also after caesarean section.

The cholinergic system, which also is of importance for vigilance, may be activated at birth. Transgenic mice lacking the β_2 subunit of the nicotinic acetylcholine receptors do not arouse to the same extent as wild-type mice (Cohen et al., 2005).

Awareness of the body and the self

Newborn infants react differently when they are touched by another person than themselves: they must have some awareness of their own bodies (Rochat, 2003). They have a proprioceptive sense of their own bodies. Thus, infants at birth do not just sense and respond to stimuli based on reflex mechanisms, but they also feel and interact with objects they experiment as differentiated to them (Rochat, 2011).

By studying the looking behaviour of newborns exposed to synchronous versus asynchronous cues it was found that they are able to perceive their bodies from birth. They can visually discriminate visual tactile stimuli when the visual information is related to the body. By using near infrared spectroscopy (NIRS), Filippetti et al. identified the same specialized areas of the temporal cortex as in adults for processing body-related information (Filippetti et al., 2014).

Pain

Pain can elicit withdrawal reflexes and release of cortisol, beta-endorphin and noradrenaline from about the twentieth gestational week. These physiological reactions are mediated at levels below the cerebral cortex. From about 25 weeks, cortical responses in the somatosensory area have been recorded by NIRS. Facial expressions similar to adults sustaining pain have been observed in preterm infants after 25 weeks. Thus these preterm infants are probably conscious about pain. However, since the foetus is exposed to high endogenous sedatory and analgesic substances it may not be conscious about pain even after 25 weeks. The foetus seems to be particularly protected from noxious stimuli during birth. Vitamin K injection given in an hour after vaginal delivery did not evoke any reaction in contrast to infants delivered by caesarean section (Bergqvist et al., 2009). This could be due to the effect of oxytocin, which hyperpolarizes the gabaergic neuronal membranes.

Smell and taste

The foetus can probably smell from about the twentieth week. It may remember certain smells and tastes it has been exposed to after birth.

Newborns seem to remember the smell of amniotic fluid (Tyzio et al., 2006), which attract them more than other cues. Behavioural responses to smell can be recorded in preterm infants from about the twenty-ninth week. Olfaction is processed in the orbitofrontal cortex which has been demonstrated with near-infrared spectroscopy (NIRS). Pleasant smells like vanilla and colostrum evoked increased hemodynamic responses (Bartocci et al., 2000) in contrast to unpleasant smells like the smell of disinfectant or detergent (Bartocci et al., 2001).

Vision

Newborn infants were earlier assumed not to be able to see or recognize anything: they only saw a fog. Although their acuity is only about 0.1, a number of studies have demonstrated that they can process complex visual stimuli, recognize and imitate facial expressions (Meltzoff and Moore, 1977). They have developed preferential looking and look longer on patterned than grey fields. Even very preterm infants seem to fix on the sight of their mothers briefly. Human newborn infants have an inborn capacity for face recognition, which is of crucial importance for the development of social networks.

Hearing

The cochlea becomes structurally developed from about the eighteenth gestational week, although the auditory does not function until after the twenty-sixth week when brain stem–evoked responses can then be recorded (Wilkinson and Jiang, 2006). The foetus can react to sound by tachycardia from the twentieth week. Cortical activation to sound was detected in the foetus from the thirty-third week. The full-term infant can orient sounds by turning the head towards the source. If they are shown an object at the same time, they will move their eyes towards the sound rather than the object, indicating that hearing is more mature than seeing at birth.

Memory

Memory is a crucial component of consciousness. Our impressions are usually related to memories. Habituation—a very short-term memory—has been demonstrated in the human foetus at around 22 to 23 weeks of gestation. Foetuses exposed to repetitive vibrations of an electric toothbrush react with movements until it habituates to the stimulus and does not react any longer (Leader et al., 1988).

Newborn infants remember sounds, melodies, and rhythmic poems they have been exposed to during foetal life (Hepper, 1996). Newborn infants can also learn a certain behaviour during sleep (Fifer et al., 2010). At about two months of age they form some kind of mental representation of faces and things (Marshall and Melzoff, 2014). Thus Piaget's concept "out of sight, out of mind" may apply only before two months and not half a year as earlier believed.

Language

Language is also an important component of human consciousness. *Infant* actually means someone without language. However, newborn infants have been found to be able to discriminate between languages belonging to different rhythmic families (Nazzi et al., 1998). Swedish newborn infants showed recognition when they were exposed to the vowel /y/ by slowly sucking a pacifier with a pressure device, while they sucked vigorously when they heard the sound /i/, less common in Swedish (Moon et al., 2013). American newborn infants behaved in the opposite way, seeming to recognize /y/ but not /i/.

Preterm infants born two to three months before term distinguished spoken syllables in a similar way as adults. This was found by measuring cerebral blood oxygenation with NIRS when exposing the infants to sounds like "ga" and "ba" and female versus male voices (Mahmoudzadeh et al., 2013).

Infants are like magnets absorbing phonemes and words. They can also distinguish between sounds in different languages. Interestingly preterm infants who have been cared at a normal busy neonatal intensive care unit with a lot of people talking to each other do better in language tests at two years of age than infants who have been isolated in private rooms. After half a year of age the ability to pick up many languages decreases.

Integration of multiple sensory modalities and memory

Infants can integrate sensory signals from different modalities to some extent. A newborn infant exposed to a nociceptive stimuli can be calmed by sucking sucrose. A four-week-old baby can connect what it sees and hears. This is in contrast to for example reptiles. A snake has no concept of a particular prey, e.g., a mouse, and no perception of object constancy. It is governed by sight to strike the prey, but to start to swallow the head of the prey, it must use its smell. They lack the "global conscious (Lagercrantz and Changeux, 2009) workspace" which develops later in evolution as well as during ontogenesis.

Emotions

Newborn human infants express so-called primordial emotions like hunger and thirst (Denton, 2005). The newborn infant is seeking for the breast areaola soon after birth. The respiratory drive or "hunger for air" which is initiated at birth can also be regarded as a primordial emotion. These primordial emotions can be regarded as marking the dawning of consciousness. An almost unique feature of the human newborn is crying that follows awakening at birth. It may reflect a sign of discomfort with the new environment. It is probably essential to mobilize the caring instincts of the parents.

But the newborn infant also expresses facial expressions of joy, particularly when it is taken up and placed at the mother's breast. Joy or hedonic feelings are important components of consciousness (Kringelbach and Berridge, 2012).

The foetus may also show emotional facial expressions of pleasure and displeasure. Whether this should be regarded as signs of consciousness can be questioned. However, using functional MRI, studies have shown that networks that make consciousness possible are already identified during the foetal life (Schöpf et al., 2012).

Social interaction

The so called "social brain" involve the temporal parietal junction, the temporal poles, and the dorsal medial prefrontal cortex (Farroni et al., 2013). By using near infrared spectroscopy (NIRS), Farroni et al. found that these areas are activated already in newborn infants (one to five days old) when exposed to dynamic face stimulus, but not by viewing human arms. This response increased during the first days after birth after the babies had been seeing many faces.

The social interaction is crucial for learning of the language. Kuhl (2010) has demonstrated that nine-month-old American infants can also learn Chinese if the instructors read and tell tales for them (Kuhl, 2010). However, this finding cannot be reproduced by using a screen, although the children are exposed to the same stories.

Already newborns imitate gestures. Reciprocal imitation games support social bonding and affiliation. The "like me" concept gives rise to a life-long ability to connect with other persons. This is vital for our survival as species.

Imitation is mediated by mirror neurons, which seem to mature early in life. Neural mirroring mechanisms establish prelinguistic mapping between self and others (Marshall and Melzoff, 2014). This includes the anticipation of emotional reactions of other people. Thus infant imitation is not an automatic uncontrolled impulse but under intentional control.

Conclusions

What it is like to be human foetus or a newborn is not only a philosophical question. It has several social and clinical implications, i.e., how we treat the foetus and the infant for example during nociceptive procedures. The foetus reacts to various sensory stimuli and is even able to remember jingles and vowels. However, it is less likely that it is conscious, since it is living at a low oxygen level immersed with inhibitory substances. Furthermore, it is mainly asleep and even if it can open its eyes it is probably not conscious. Memories seem to be preserved during REM-sleep, which is the predominate state of the foetus.

The newborn infant is awakened due to the activation of the sympathoadrenal system in the body and locus coerules in the brain. It is aware of itself and the environment, expresses emotions and joy. There is a biophysical indication of consciousness by the observation of a limited default mode network.

Even the preterm infant ex utero may open its eyes and establish a minimal eye contact with its mother and show other signs of consciousness like cortical responses to pain. However, if the preterm infant is born before 23 to

25 weeks' gestation, it is less likely to be conscious. The input from the sensory organs via the thalamocortical connections have not yet been established which is required for awareness.

References

Bartocci, M., Bergqvist, L. L., Lagercrantz, H., & Anand, K. J. 2006. Pain activates cortical areas in the preterm newborn brain. *Pain*, 122, 109–17.

Bartocci, M., Winberg, J., Papendieck, G., Mustica, T., Serra, G., & Lagercrantz, H. 2001. Cerebral hemodynamic response to unpleasant odors in the preterm newborn measured by near-infrared spectroscopy. *Pediatr Res*, 50, 324–30.

Bartocci, M., Winberg, J., Ruggiero, C., Bergqvist, L. L., Serra, G., & Lagercrantz, H. 2000. Activation of olfactory cortex in newborn infants after odor stimulation: A functional near-infrared spectroscopy study. *Pediatr Res*, 48, 18–23.

Bartra, R. 2014. *Anthropology of the brain*. Cambridge, UK: Cambridge University Press.

Barttfeld, P., Uhrig, L., Sitt, J. D., Sigman, M., Jarraya, B., & Dehaene, S. 2015. Signature of consciousness in the dynamics of resting-state brain activity. *PNAS*, 112, 887–92.

Bergqvist, L. L., Katz-Salamon, M., Hertegard, S., Anand, K. J., & Lagercrantz, H. 2009. Mode of delivery modulates physiological and behavioral responses to neonatal pain. *J Perinatol*, 29, 44–50.

Born, P., Rostrup, E., Leth, H., & Al, E. 1996. Change of visually induced cortical activation patterns during development. *Lancet*, 347, 543.

Bourgeiois, J. P. 2010. The neonatal synaptic big bang. In Lagercrantz, H., Hanson, M. A., Ment, L. R., & Peebles, D. M. (eds.), *The newborn brain*. 2nd ed. Cambridge, UK: Cambridge University Press.

Byrge, L., Sporns, O., & Smith, L. B. 2014. Developmental process emerges from extended brain-body-behavior networks. *Trends Cogn Sci*, 18, 395–403.

Chalmers, D. 1996. *The conscious mind. In search of a fundamental theory*. Oxford, UK: Oxford University Press.

Changeux, J. P., & Dehaene, S. 1989. Neuronal models of cognitive functions. *Cognition*, 33, 63–109.

Cohen, G., Roux, J. C., Grailhe, R., Malcolm, G., Changeux, J. P., & Lagercrantz, H. 2005. Perinatal exposure to nicotine causes deficits associated with a loss of nicotinic receptor function. *Proc Natl Acad Sci USA*, 102, 3817–21.

Cotterill, R. 2003. Cyber child: Aimulation test-bed for consciousness studies. *J Consciousness Stud*, 10, 31–45.

Deco, G., Hagmann, P., Hudetz, A. G., & Tononi, G. 2014. Modeling resting-state functional networks when the cortex falls asleep: Local and global changes. *Cereb Cortex*, 24, 3180–94.

Dehaene, S. 2014. *Consciousness and the brain deciphering how the brain codes our thoughts*. New York: Viking.

Demertzi, A., Gomez, F., Crone, J. S., Vanhaudenhuyse, A., Tshibanda, L., Noirhomme, Q., Thonnard, M., Charland-Verville, V., Kirsch, M., Laureys, S., & Soddu, A. 2014. Multiple fMRI system-level baseline connectivity is disrupted in patients with consciousness alterations. *Cortex*, 52, 35–46.

Denton, D. 2005. *The primordial emotions*., Oxford, UK: Oxford University Press.

Doria, V., Beckmann, C. F., Arichi, T., Merchant, N., Groppo, M., Turkheimer, F. E., Counsell, S. J., Murgasova, M., Aljabar, P., Nunes, R. G., Larkman, D. J., Rees, G., & Edwards, A. D. 2010. Emergence of resting state networks in the preterm human brain. *PNAS*, 107, 20015–20.

Farroni, T., Chiarelli, A. M., Lloyd-Fox, S., Massaccesi, S., Merla, A., Di Gangi, V., Mattarello, T., Faraguna, D., & Johnson, M. H. 2013. Infant cortex responds to other humans from shortly after birth. *Sci Rep*, 3.

Fifer, W. P., Byrd, D. L., Kaku, M., Eigsti, I.- M., Isler, J. R., Grose-Fifer, J., Tarullo, A. R., Balsam, P. D. 2010. Newborn infants learn during sleep. *PNAS*, 107, 10320–3.

Filippetti, M. L., Lloyd-Fox, S., Longo, M. R., Farroni, T., & Johnson, M. H. 2014. Neural mechanisms of body awareness in infants. *Cereb Cortex*, 1–9.

Fransson, P., Metsäranta, M., Blennow, M., Åden, U., Lagercrantz, H., & Vanhatalo, S. 2013. Early development of spatial patterns of power-law frequency scaling in fMRI resting-state and EEG data in the newborn brain. *Cereb Cortex*, 23, 838–46.

Fransson, P., Aden, U., Blennow, M., & Lagercrantz, H. 2011. The functional architecture of the infant brain as revealed by resting-state fMRI. *Cereb Cortex*, 21, 145–54.

Fransson, P., Skiold, B., Horsch, S., Nordell, A., Blennow, M., Lagercrantz, H., & Aden, U. 2007. Resting-state networks in the infant brain. *Proc Natl Acad Sci USA*, 104, 15531–6.

Fransson, P., Skiold, B., Engstrom, M., Hallberg, B., Mosskin, M., Aden, U., Lagercrantz, H., & Blennow, M. 2009. Spontaneous brain activity in the newborn brain during natural sleep—an fMRI study in infants born at full term. *Pediatr Res*, 66, 301–5.

Gao, W., Zhu, H., Giovanello, K. S., Smith, J. K., Shen, D., Gilmore, J. H., & Lin, W. 2009. Evidence on the emergence of the brain's default network from 2-week-old to 2-year-old healthy pediatric subjects. *Proc Natl Acad Sci USA*, 106, 6790–5.

Gluckman, P. E. A. 1983. The effect of cooling on breathing and shivering in unanaestetized foetal lambs in utero. *J Physiol*, 343, 495.

Han, S., & Northoff, G. 2008. Culture-sensitive neural substrates of human cognition: A transcultural neuroimaging approach. *Nat Rev Neurosci*, 9, 646–54.

He, B. J., & Raichle, M. E. 2009. The fMRI signal, slow cortical potential, and consciousness. *Trends Cogn Sci*, 13, 302–9.

Hepper, P. 1996. Fetal memory: Does it exist? What does it do? *Acta Paediatrica*, Suppl. 416, 16–20.

Heuvel, M. P. Van Den, & Sporns, O. 2011. Rich-club organization of the human connectome. *J Neurosci*, 31, 15775–86.

Koch, C. 2004. *The quest for consciousness: A neurobiological approach.* Eaglewood, Colorado: Roberts & Company Publishers.

Kostovic, I., & Judas, M. 2010. The development of the subplate and thalamocortical connections in the human foetal brain. *Acta Paediatr*, 99, 1119–27.

Kouider, S., Stahlhut, C., Geiskov, S. V., Barbosa, L. S., Dutat, M., De Gardelle, V., Christophe, A., Dehaene, S., & Dehaene-Lambertz, G. 2013. A neural marker of perceptual consciousness in infants. *Science*, 340, 376–80.

Kringelbach, M. L., & Berridge, K. C. 2012. The joyful mind. *Sci Am*, 307, 40–5.

Kuhl, P. 2010. Brain mechanisms in early language acquisition. *Neuron*, 67, 713–27.

Lagercrantz, H. 1996. Stress, arousal, and gene activation at birth. *New Physiol Sc* 214–18.

Lagercrantz, H. 2014. The emergence of consciousness. *Semin Fetal Neonat M*, 19, 300–5.

Lagercrantz, H., & Changeux, J. P. 2009. The emergence of human consciousness: from fetal to neonatal life. *Pediatr Res*, 65, 255–60.

Lagercrantz, H., Hanson, M., Ment, L., & Rodeck, C. (EDS.) 2010. *The newborn brain.* Cambridge: Cambridge University Press.

Lagercrantz, H., Pequignot, J., Pequignot, J. M., & Peyrin, L. 1992. The first breaths of air stimulate noradrenaline turnover in the brain of the newborn rat. *Acta Physiol Scand*, 144, 433–8.

Lagercrantz, H., & Slotkin, T. A. 1986. The "stress" of being born. *Sci Am*, 254, 100–7.

Laureys, S., & Tononi, G. 2008. *The neurology of consciousness*. London: Academic Press.

Leader, L. R., Stevens, A. D., & Lumbers, E. R. 1988. Measurement of fetal responses to vibroacoustic stimuli. Habituation in fetal sheep. *Biol Neonate*, 53, 73–85.

Lombardo, M. V., Chakrabarti, B., Bullmore, E. T., Sadek, S. A., Pasco, G., Wheelwright, S. J., Suckling, J., & Baron-Cohen, S. 2010. Atypical neural self-representation in autism. *Brain*, 133, 611–24.

Mahmoudzadeh, M., Dehaene-Lambertz, G., Fournier, M., Kongolo, G., Goudjil, S., Dubois, J., Grebe, R., & Wallois, F. 2013. Syllabic discrimination in premature human infants prior to complete formation of cortical layers. *PNAS*, 110, 4846–51.

Marshall, P. J., & Melzoff, A. M. 2014. Neural mirroring mechanisms and imitation in human infants. *Philos T Roy Soc B*, 369.

Mellor, D. J., Diesch, T. J., Gunn, A. J., & Bennet, L. 2005. The importance of awareness for understanding fetal pain. *Brain Res Brain Res Rev*, 49, 455–71.

Meltzoff, A. N., & Moore, M. K. 1977. Imitation of facial and manual gestures by human neonates. *Science*, 198, 75–8.

Moon, C., Lagercrantz, H., & Kuhl, P. 2013. Language experienced in utero affects vowel perception after birth: A two-country study. *Acta Paediatr*, 102, 158–60.

Nagel, T. 1974. What is it like to be a bat? *Philos Rev*, 83, 435–50.

Nazzi, T., Bertoncini, J., & Mehler, J. 1998. Language discrimination by newborns: Toward an understanding of the role of rhythm. *J Exp Psychol Hum Percept Perform*, 24, 756–66.

Northoff, G. 2010. Humans, brains, and their environment: Marriage between neuroscience and anthropology? *Neuron*, 65, 748–51.

Nowakowsky, R. S. 2006. Stable neuron numbers from cradle to grave. *PNAS*, 103, 12219–20.

Posner, M. I., & Rothbart, M. K. 1998. Attention, self-regulation, and consciousness. *Phil Trans R Soc Lond B*, 353, 1915–27.

Rochat, P. 2003. Five levels of self-awareness as they unfold early in life. *Conscious Cogn*, 12, 717–31.

Rochat, P. 2011. The self as phenotype. *Conscious Cogn*, 20, 100–19.

Schleger, F., Landerl, K., Muenssinger, J., & Al, E. 2014. Magnetoencephalographic signatures of numerosity discrimination in fetuses and neonates. *Dev Neuropsychol*, 39, 316–29.

Schöpf, V., Kasprian, G., Brugger, P. C., & Prayer, D. 2012. Watching the fetal brain at "rest." *Int J Devl Neuroscience*, 30, 11–17.

Searle, J. R. 2000. Consciousness. *Annu Rev Neurosci*, 23, 557–78.

Thomason, M. E., Brown, J. A., Dassanayake, M. T., Shastri, R., Marusak, H. A., Hernandez-Andrade, E., Yeo, L., Mody, S., Berman, S., Hassan, S. S., & Romero, R. 2014. Intrinsic functional brain architecture derived from graph theoretical analysis in the human fetus. *PLoS One*, 9, e94423.

Tyzio, R., Cossart, R., Khalilov, I., Minlebaev, M., Hubner, C. A., Represa, A., Ben-Ari, Y., & Khazipov, R. 2006. Maternal oxytocin triggers a transient inhibitory switch in GABA signaling in the fetal brain during delivery. *Science*, 314, 1788–92.

Vanhatalo, S., & Kaila, K. 2006. Development of neonatal EEG activity: From phenomenology to physiology. *Semin Fetal Neonatal Med*, 11, 471–8.

Vanhatalo, S., & Lauronen, L. 2006. Neonatal SEP – back to bedside with basic science. *Semin Fetal Neonatal Med*, 11, 464–70.

Vanhaudenhuyse, A., Noirhomme, Q., Tshibanda, L. J., Bruno, M. A., Boveroux, P., Schnakers, C., Soddu, A., Perlbarg, V., Ledoux, D., Brichant, J. F., Moonen, G., Maquet, P., Greicius, M. D., Laureys, S., & Boly, M. 2010. Default network connectivity reflects the level of consciousness in non-communicative brain-damaged patients. *Brain*, 133, 161–71.

Wilkinson, A. R., & Jiang, Z. D. 2006. Brain stem auditory evoked response in neonatal neurology. *Semin Fetal Neonatal Med*, 11, 444–51.

Zelazo, P. D. 2004. The development of conscious control in childhood. *Trends Cogn Sci*, 8, 12–17.

2 Mapping mind-brain development

Andreas Demetriou, George Spanoudis, and Michael Shayer

It is taken for granted that the brain is the underlying biological mechanism of the mind, because the mind emerges, in all of its expressions, from the structure and functioning of the brain. We aim to examine how these levels of analysis, the psychological and the biological, interact with each other. Thus, we intend to critically review the principles underlying the brain's capability to generate mind and the mind's potential to direct the functioning of the brain. The frame for discussion is our theory of intellectual development. Here we will only outline the basic postulates of the theory (for more details see Demetriou et al., 2010; Demetriou et al., 2014a).

The developing mind

Architecture

The human mind is organized into systems carrying out different tasks during understanding or problem solving. There are four types of systems: *domain-specific, representational, integrative*, and *cognizance systems*. Domain-specific systems are directly related to the environment, grounding the mind in the real world. The other three types of systems are increasingly detached from the environment, organizing environment-specific information, mental constructs, and behavior at various levels of abstraction. This architecture, which is illustrated in Figure 2.1, is summarized below. It is stressed that systems and processes in this architecture are identifiable as distinct entities by various methods, such as structural equation modeling (Demetriou et al., 2010).

Domains of mental functioning. Research suggests the following domains of thought: Categorical (e.g., physical similarity in color, shape), spatial (e.g., arrangement in space, such as close and far away, left-right), quantitative (e.g., aggregation and distribution of objects), causal, (e.g., effective interactions), and social thought (e.g., species-specific important information, such as face of conspecifics, comforting behaviors). Each of these domains specializes in the processing of specific relations that are so important for routine functioning in the environment that they are seeded into the perceptual systems themselves.

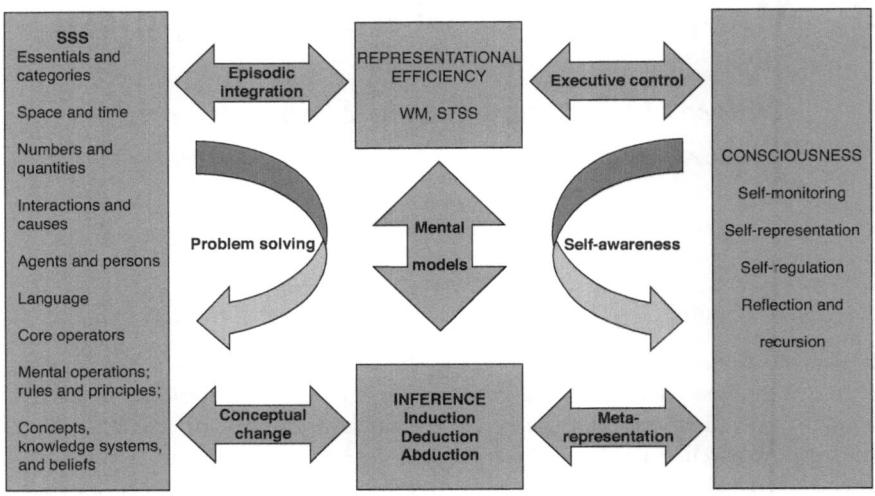

Figure 2.1 General architecture of the human mind.

Note: SSS = Specialized Structural Systems. WM = Working Memory. STSS = Short-term Storage Space.

Each domain involves (i) core processes (e.g., color perception, depth perception, subitization), (ii) mental operations (e.g., sorting, mental rotation, numerical operations), and (iii) knowledge and beliefs (e.g., categories about objects, mental maps of a city) (Demetriou et al., 2010).

Representational efficiency. To recognize stimuli, note similarities and differences between them, and reason about them, the mind requires a workspace where current information may be represented and processed. The capacity of this workspace, commonly known as working memory, relates to the kind and complexity of the relations that may be worked out. It involves: (i) modality-specific storage systems, such as acoustic and visual storage; (ii) a central executive, which represents the currently active mental goal and directs further information search; and (iii) an episodic buffer integrating information across modes into unitary episodic representations preserving time and the subjectivity of experience (Baddeley, 2012).

Integrative processes. Episodic integration is the starting point of integration geared in the ongoing flow of experience. However, meaning making often requires integration of information in working memory with past encounters and knowledge. This is carried out by a triple-process mechanism: Abstraction, Alignment, and Cognizance, the AACog mechanism. Abstraction extracts similarities between representations, alignment interlinks them, and cognizance reduces them into new representations (Demetriou and Kazi, 2006; Demetriou et al., 2014a). AACog is equivalent to psychometric general intelligence, the well-known g factor (Jensen, 1998). *Inference*, inductive and deductive, is a fundamental tool of this mechanism because it carries properties across representations and ensures their evaluation.

Consciousness. Consciousness is an invention of evolution for the sake of monitoring the ongoing flow of information processing and extending experience beyond the here and now. It reflects the mind's need to rework experiences in order to select among them, stabilize whatever appropriate, and integrate with past experience. Cognizance is the component of consciousness focusing on the mind itself. So defined, cognizance allows feedback loops where cycles of abstraction, alignment, and inference may become the object of further abstraction and alignment. Metarepresentation is a fundamental component of cognizance in that it encodes similarities between representations into new representations (Demetriou et al., 2014a).

Development

AACog is active in its entirety at birth. However, the operation of each function varies with development. AACog evolves through four major developmental cycles: episodic representations, mental representations, rule-based representations, and principle-based representations (Demetriou et al., 2014a). Each cycle involves two phases: production of new mental units and alignment. The production of a new kind of representation dominates in the first phase of each period (i.e., gross episodic representations, gross representational blocks, generic rules, and general principles from 0–1, 2–4, 6–8, and 11–13 years, respectively), when abstraction process dominates. The integration of the cycle-specific representations into more complex ensembles dominates in the second (i.e., integrated episodic sequences at 1–2, representational alignments at 4–6, conceptual alignment at 8–11, and principle alignment at 13–16 years, respectively). In this phase alignment process dominates. Therefore, in each period, representations proliferate in the first phase and they are integrated with each other in the second phase, enabling transition to the next cycle. Cognizance about the representational units starts low at the beginning and culminates at the end of each cycle. That is, awareness about the cycle-specific representations displaces processing from the current mental constructs to their underlying relations, which may be metarepresented into the constructs of the next period (Demetriou et al., 2014a; Spanoudis et al., 2015).

The relations between this sequence of cycles and important factors of representational efficiency, such as processing speed and working memory, vary systematically. We showed that changes at the first phase of each cycle (i.e., at 2–4 years, 6–8 years, and 11–13 years) are predicted by changes in processing speed while changes at the second phase (i.e., 4–6 years, 8–10 years, and 14–16 years) are predicted by changes in working memory. These changes reflect differences in the processing requirements of developmental acquisitions. At the beginning of cycles, processing speed is a better index because thought in terms of the new mental units gets automatized and expands fast over different contents. Later in the cycle, working memory is a better index because alignment and inter-linking of representations both requires and facilitates working memory. However, speed and working memory index rather than cause transitions from one cycle to another. Cognizance is the main causal factor of transition across cycles because it renders cognitive experiences available to abstraction and alignment, thereby producing new mental content that may be metarepresented into new mental units (Demetriou et al., 2013; Demetriou et al., 2014b).

The four cycles may be described from several aspects. Here we will focus on executive control and working memory because they are central to current research on the brain. Although limited and perception- and action-based, executive control is present in the first cycle. Selection of an action plan between alternatives indicates a minimum of awareness and planning that is integral to executive control. For example, infants in their early second year imitate actions involving parts of their body that are not visible to them, such as the head. The information that can be held in working memory in each developmental cycle relates to the dominant executive control programs of the cycle. Infants can represent in working memory information about one object for several seconds as of the age of 6 months (Kaldy and Leslie, 2005). Working memory for actions develops systematically throughout infancy (Morra et al., 2014).

Executive control in the preschool years, at 3–4 years, is expressed as a *representational control program* allowing toddlers to focus on two to three interrelated representations and alternate between them while both are in focus. Technically, this program is represented by various tasks (e.g., go/no go tasks) requiring from the child to inhibit responding to one perceptually strong stimulus in order to respond to a goal-relevant stimulus masked by the strong stimulus (Demetriou et al., 2014a). Visual working memory develops extensively in this period, developing from about one element at the age of 3 to about four elements at the age of 8 years. Verbal working memory is limited to one to two elements throughout this period (Demetriou et al., 2013).

In the next cycle executive control is upgraded into *a conceptual fluency program* allowing children to shift between conceptual spaces (e.g., various object categories), activate space-specific instances, and interrelate them according to specific conceptual constraints. For example, children at 8–9 years of age may perform well on tasks requiring shift between conceptual spaces by recalling words starting with particular letters (Demetriou et al., 2014a, 2014b). Working memory turns verbal in this period. Verbal working memory develops from two elements at the age of 6–7 years to more than four elements at the age of 12 years (Demetriou et al., 2013). Pickering (2001) showed that recoding of visually presented information into a phonological form appears at the age of 8 years, which coincides with the shift from visual to verbal working memory development. This is upgraded into *an inferential relevance mastery program* in adolescence but this program will not be analyzed here for space considerations.

Language of thought

Is there a language of thought (LOT)? That is, are there standard mental units and mental rules underlying the transformation and combination of mental units (Fodor, 1975)? The model above implies that a LOT does exist and it is as an emergent aspect of intellectual development, involving mental units and rules that are reformed in development. At birth, mental units come from the core operators associated with the various domain-specific systems (e.g., color based categories, subitization-based quantities). Syntactic rules capture environment-based variations of these mental units (e.g., color change distancing an object from an X-color-based category antedates sorting, addition or removal of an object from a subitized set antedates numerical operations). Thus, initial meaning about the world emerges from perception-based standard relations in the world.

AACog is the central operating system underlying LOT expansion. Combinations between core units within and across domain-specific systems are framed and streamed by the three fundamental AACog processes. These combinations generate new mental objects organizing simpler input representations, such as class hierarchies (e.g., cat, pet, animal) or the mental number line. The relations between representations within and across levels of mental hierarchies are specified by inferential processes, such as inductive and deductive inference, that ensure truthful and valid productivity in the inclusion of new instances in the system or generation of implied instances. Class hierarchies and sorting in categorical thought, the mental number line and arithmetic in quantitative thought, mental maps of the environment, and visuo-spatial operations in the spatial system, are examples of the mental objects and rule systems in this LOT. Specific rules in each of these systems are implementations of general inferential rules that may be called upon to evaluate cohesion and consistency, such as conjunction, disjunction, modus ponens, etc.

Cognizance is a major contributor to LOT expansion, recycling along the mental units and processes of each cycle (Demetriou et al., 2010, p. 329). Specifically, in the first cycle, in the second year of life, infants show an awareness of their body and their actions (Demetriou et al., 2014a). At the beginning of the second cycle, toddlers show an awareness of perception as a source of knowledge; at 4–5 years they become aware of their own and other persons' representations. In the next cycle, at 8–9 years, children show awareness of the differences between mental functions, such as perception and memory and the underlying mental processes connecting representations, such as syntax in language or inference in reasoning (Spanoudis et al., 2015). In the next cycle, at 13–14 years, adolescents are explicitly aware of logical relations and principles and they accurately represent mental functions and domains (Demetriou and Kazi, 2006; Demetriou et al., 2014a).

The developing brain

Mapping the developing mind-brain relations

What are the neuronal underpinnings of the mental architecture and development outlined above? This may be answered by answering the following questions:

1 Is the architecture of mind outlined above reflected in the organization and functioning of the brain?
2 What are the neuronal analogues of the various mental processes and mechanisms? For instance, how are abstraction, alignment, and cognizance carried out in the brain?

3 How are brain changes (such as increases in neuronal volume, myelination, neuronal networking, and neuronal pruning) related to the phase and cycle changes of cognitive processes?

We caution that answering these questions is not an easy matter. On the one hand, psychological research involves: (i) observable responses expressed in different scales, and (ii) subjective experiences expressed in various modes. On the other hand, brain research involves biological entities, such as the nature, volume, and organization of neuronal matter and its functional correlates, expressed through various modes (e.g., blood supply and glucose consumption of activated brain areas and electrical activity). The scales and precision of measuring each one of these dimensions vary both within and across the two levels of description. Thus, we do not yet know how to map the different levels of analysis onto each other and when we map them precision falls short of the optimal.

Mind-related brain architecture

Domain-specific networks

Each of the relational domains uncovered by psychological research may be mapped onto one or more specifically dedicated brain structure or networks. For instance, categorization of properties is processed in the visual cortex (Galaburda, 2002), visuo-spatial relations such as mental rotation are processed in the intraparietal sulcus (Harris et al., 2000), quantitative information is processed in the inferior parietal cortex (BA 39) (Dehaene, 2011), understanding of causal interactions is processed in the medial and dorsal part of the superior frontal cortex (Fonlupt, 2003), and face recognition involves the fusiform gyrus (Kanwisher et al., 1997). Admittedly, there is not as yet a full map of the networks serving each of the domains at its three levels of organization (i.e., core processes, mental operations and belief systems, at different age phases). Table 2.1 shows brain regions associated with mental processes and Figure 2.2 localizes the processes in the brain.

Table 2.1 Locations of mental functions in the brain

No. in Fig. 2	Mental function	Brain region
	Domain-specific	
1	Categorical	Occipital, superior temporal gyrus
2	Spatial	Occipital
3	Quantitative	Inferior parietal (angular gyrus (BA 39)
4	Causal	Occipital, superior frontal (medial and dorsal)
5	Social (face recognition)	Medial prefrontal, superior temporal sulcus, temporal poles

No. in Fig. 2	Mental function	Brain region
	STS	
6	Visual	Posterior parts of the superior frontal sulcus, entire interparietal sulcus, amygdala, hippocampus
7	Verbal, rehearsal	Left-lateralized premotor-parietal
8	Verbal, maintenance	Anterior-prefrontal/inferior parietal
9	Episodic	Right middle frontal gyrus, pre-supplementary motor area (pre-SMA)
	Executive control	
10	Inhibition	Ventral and dorsal prefrontal
11	Selection	Medial pre-frontal, parieto-temporal association areas
12	Switching	The inferior frontal junction, premotor-intraparietal network
	Reasoning	
13	Binding	Hippocampus, medial temporal and inferior temporal, dorsal and/or anterior PFC
14	Inductive	Inferior frontal gyrus, right insular cortex
15	Deductive	Temporal (BA 21, 22), frontal regions (BA 44, 8, 9), occipital (BA 18, 19), left parietal (BA 40), bilateral dorsal frontal (BA 6), left frontal (BA 44, 8, 10), right frontal (BA 46), right superior parietal lobule, thalamus, right anterior cingulate
	Consciousness	
16	Visual	Claustrum
17	Theory of mind	Right and left temporo-parietal junction, medial parietal cortex (including posterior cingulate and precuneus), and medial prefrontal cortex (mPFC), ventral and dorsal attentional

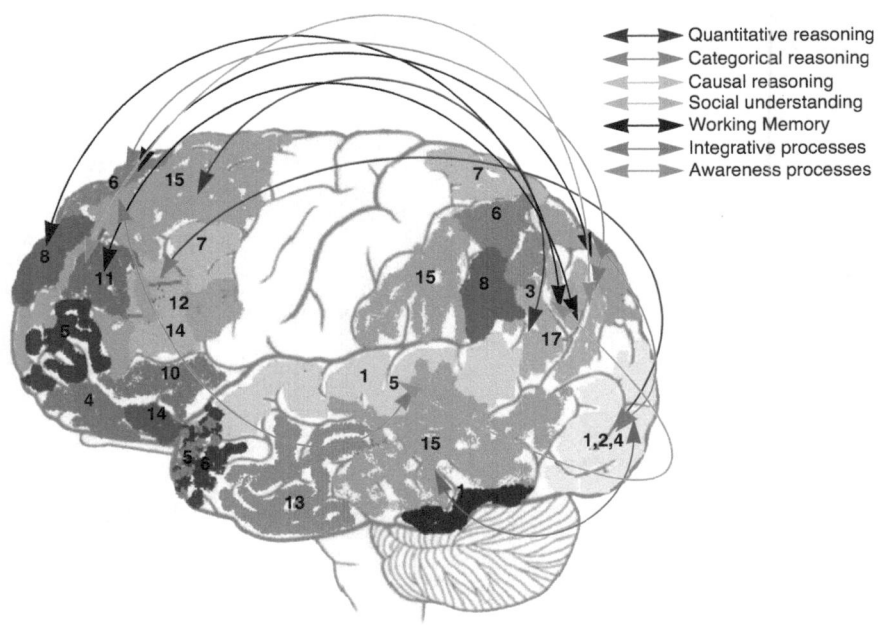

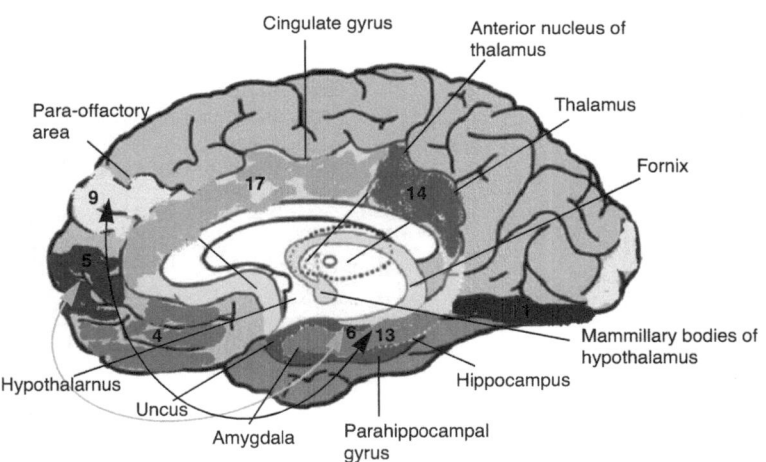

Figure 2.2 Brain regions associated with mental processes presented in Table 2.1.
A color version of this image is available here: www.routledge.com/
products/9781138851672.

Networks underlying representational efficiency

It is also clear that several structures and networks serve working memory. First, verbal and visual short-term storage are served by different networks. Specifically, verbal storage is served by two distinct circuits, one specializing in verbal rehearsal as such (i.e., a left-lateralized premotor-parietal network) and another one serving nonarticulatory maintenance of phonological information (i.e., a bilateral anterior-prefrontal/inferior parietal network). Visuo-spatial storage relies on only one clearly different bilateral brain system: The posterior parts of the superior frontal sulcus and the entire interparietal sulcus. Some brain regions (i.e., right middle frontal gyrus and the pre-SMA as well as bilaterally in the deep frontal opercular cortex and the cortex along anterior and middle parts of the intraparietal sulcus) are activated during processing of both, the verbal and the visuo-spatial tasks. This is in line with the assumption that a central episodic buffer may exist (Baddeley, 2012; Repovš and Baddeley, 2006).

Hippocampus is crucial for several aspects of working memory, especially spatial working memory (Squire, 1992). It is involved in the explicit representation of information rather than in implicitly represented memories related to automated skills and habits. It temporarily binds together several distributed areas in the cortex representing various kinds of information. In fact, it is itself differentiated so that different structures in it serve different aspects of working memory. Specifically, amygdala encodes information to be retained in working memory and CA encodes rules specifying spatial arrangements, such as object location, and mediates with the frontal cortex (Friedman and Goldman-Rakic, 1988). This mediation is necessary because working memory functions are widely distributed: hippocampus represents novel information, medial temporal and inferior temporal regions represent visual and multimodal object representations, dorsal and/ or anterior PFC capture relations between objects, and ventral and/or posterior PFC maintains the currently relevant items (Raganath and D' Esposito, 2005). The Inferior Frontal Junction (IFJ), which is situated at the junction of the precentral sulcus and the inferior frontal sulcus, carries task switching, causing the attentional bottleneck because it can retain only one rule at a time (Vergauwe et al., 2015). Finally, response selection is carried out by medial pre-frontal regions (Cowey, 1996; Gruber and Goschke, 2004). Thus the executive functions of working memory are carried out in the frontal cortex rather than in the hippocampus as such.

Networks underlying integrative inferential processes

Obviously, the executive networks above are part of the integrative system of the brain. These systems are also activated by reasoning together with regions serving their specific requirements. Specifically, inductive reasoning activates the inferior frontal gyrus (serving integration) and the right insular cortex (serving salience detection and switching between large scale networks to set attention and working memory resources in the service of the salient representation selected;

Menon and Uddin, 2010). Deductive reasoning activates a set of networks serving different tasks at different stages of the inferential process. Specifically, content-based propositions activate temporal (BA 21, 22, serving language processing) and frontal regions (BA 44, 8, 9, serving integration). Formal propositions activate occipital (BA 18, 19, suggesting the construction of visual mental models of the relations implied by the formal propositions), left parietal (BA 40, building associations), and bilateral dorsal frontal (BA 6), left frontal (BA 44, 8, 10), and right frontal (BA 46) regions, serving integration, evaluation and selection (Goel et al., 2000).

Jung and Haier (2007) advanced an integrated model of brain functioning, the parieto-frontal integration theory (P-FIT), claiming that it is the brain analogue of fluid intelligence. Here we will map the P-FIT model onto the AACog mechanism. According to P-FIT, information is first registered and processed in regions of the cortex which specializes in dealing with different types of sensory information, such as the visual (BA 18, 19) and the auditory cortex (BA 22). Therefore, the sensory areas involved in the P-FIT model may be more related to the domain-specific processes represented in the present model, such as categorical, spatial, or quantitative core operations as noted above. From there, information is fed forward to several regions in the parietal cortex (BA 7, 39, 40), which primarily carry out elaboration, association with past knowledge or action, and abstraction. Therefore, the parietal areas of the P-FIT model may be related to the abstraction and alignment processes involved in AACog. Then these regions interact with frontal regions (BA 6, 9, 10, 45–47) in search of alternative solutions to the problem at hand. Finally, the anterior cingulate (BA 32) is engaged to constrain response selection by inhibiting alternative responses. The anterior cingulate is supposed to intervene as a conductor orchestrating when each network is to come into play in the sequence of brain events needed to reach a final decision. The anterior cingulate is called upon when there is interference caused by the fact that the same neural networks are activated by different blocks of information (Gruber and von Cramon, 2003; Klingberg, 1998). Therefore, the frontal areas of the P-FIT model relate to working memory, attention, and executive control; the anterior cingulate relates to intentional planning, inhibition, and selection, which require cognizance and metarepresentation.

Networks underlying awareness and consciousness

There is no consensus about the brain bases of consciousness. Several scholars maintain that consciousness is associated with specifically dedicated brain structures (Crick and Koch, 2005). Others maintain that consciousness emerges from the recurrent activation of the same network or the coordinated activation of complementary networks (Edelman and Tononi, 2000). The attention (controlling alerting and orienting) and executive control (handling conflicts) networks and their interplay with other networks are strong candidates for contributing to the emergence of consciousness. Alerting and orienting may be the gate and executive control the booster of awareness (Petersen and Posner, 2012).

This is suggested by students of the brain bases of the theory of mind who look for a mentalizing network serving awareness of mental states. Research suggests that this network includes right and left temporo-parietal junction (TPJ), medial parietal cortex (including posterior cingulate and precuneus), and medial pre-frontal cortex (mPFC) (Siegal and Varley, 2002). Other studies showed that ventral and dorsal attentional systems are activated by the mentalizing network to regulate the processing of self and other mental states (Abu-Akel and Shamay-Tsoory, 2011). Obviously, mentalizing ability and executive control are served by the same functional and neural systems.

How may awareness arise? According to the global workspace model advanced by Baars (1989), many modular cerebral networks are always active, unconsciously processing information in parallel. The content of processing will become the object of explicit awareness when the corresponding neural population is mobilized by top-down attentional amplification into a self-sustained brain-scale state of coherent activity that involves many neurons distributed throughout the brain. It is as though this particular content seizes the whole brain for some time (Sergent and Naccache, 2012, p. 102).

The language of the brain (LOB)

So far, we talked about interactions between brain networks related to cognitive processes but we did not specify the code of these interactions. It seems that this code is based on the various rhythms of brain activation. Brain rhythms are periodic oscillations in excitability of groups of neurons as reflected in EEG activity. Rhythms vary from very low (i.e., .05 Hz frequency) to very high (200–600 Hz). Perceptual and cognitive activity is mainly expressed into theta (4–10 Hz), beta (10–30 Hz), and gamma rhythms (30–80 Hz). It seems that components of stimuli, such as successive letters or digits presented in working memory experiments, are encoded by high frequency rhythms, such as gamma oscillations. These stand for the neural letters of thought. These letters are combined into "neural words" and neural sentences according to a specific rule, such as their presentation order, are encoded by lower frequency rhythms, such as theta oscillations. According to Buzsaki and Brendon (2012), these rhythms constitute the basic components and syntactic rules of brain language.

Interestingly, the dialogue between regions serving working memory functions in the hippocampus, such as place order of objects, and regions in the prefrontal cortex serving executive functions is held as a sequence of gamma oscillations representing items in the hippocampus and theta oscillations in the prefrontal cortex. These latter theta oscillations organize hippocampal gamma oscillations into the proper sequence. It has been suggested that working memory capacity equals the number of gamma cycles (standing for individual stimuli to be stored) that can go into a theta cycle. Thus, integrated gamma/theta cycles stand for a brain code for storing multiple items in working memory (Lisman, 2005). It has recently been suggested that theta activity is the fundamental integrative mechanism of the brain that coordinates different types of information expressed into other brain rhythms (Sauseng et al., 2010). Therefore, increasing coordination between brain rhythms in the LOB may underlie the coordination between inferential schemes underlying the emergence of LOT.

Mind-related brain development

It is well established that the brain undergoes both structural and functional changes in development. These changes reflect a dynamic interplay of simultaneously occurring progressive events, such as increases in neuronal matter and synapses, and regressive events, such as cell death and synaptic pruning. Although the total brain size is about 90 percent of adult size by age 6 years, the brain continues to undergo dynamic changes throughout adolescence and well into young adulthood. Connections between neurons also change systematically in development, although synaptogenesis and synaptic pruning differ across brain regions. Specifically, these changes peak at the second month after birth in the sensorimotor cortex, at the end of the first year in the parietal and the temporal association cortices, and from 4 to 6 years in the prefrontal cortex (e.g., Somsen et al., 1997).

Cycles in brain development

There is no research explicitly connecting the alternating cycles observed here with differential patterns in brain development. However, there are brain changes that may be related. Specifically, on the one hand, there is strong evidence that the foundations of resting-state networks are already in place before birth, with rapid neural growth in the last three months of pregnancy (Hoff et al., 2013). These networks include bilateral primary motor, primary visual and extra-striate visual, parietal-frontal and frontal (executive control networks) insular-temporal/ACC (salience and shifting networks). It is notable that, although in place, these networks reach maturity levels at successive age phases which correspond to the changes in symbolic units that dominate in working memory in each cycle (Hoff et al., 2013). Early in infancy, basic acoustic, visual, and motor networks are active, providing the basis for episodic mental units (Franson et al., 2011). These areas are connected, but minimally, with the integrative superior parietal cortex. These connections are strengthened throughout childhood, from 3 to 7 years, probably explicating the spurt in visual working memory and related mental activity. However, these connections are still weaker than in adulthood, reaching maturity in middle adolescence. Interestingly, in adolescence long distance connections between prefrontal hubs and the cerebellum are established, which may relate to fine-tuning and error detection by spotting network (and representational) inconsistencies (Hwang et al., 2012). Along the same lines, several studies showed similar cycles in interhemispheric connectivity as reflected in increases in fibers in the corpus callosum connecting the two hemispheres. That is, these fibers first increase in the frontal circuits of the corpus callosum, which sustain vigilance and regulate action planning and later, in the 6–7 years phase, there is dramatic change in the callosal isthmus, which supports interparietal associative and language functions. Later, in the 11–15 years period, these changes continue but they are attenuated. Thus, it seems that structural potentials of the brain for the episodic (primarily sensorimotor), the realistic mental (primarily visual), and the rule-based (primarily verbal) mental representations are made available at the beginning of the respective cycles.

Interestingly, these changes are tractable in spurts in the amount of EEG energy found in the alpha-frequencies. Specifically, Thatcher (1992, 1994) found that electroencephalographic coherency, which reflects changes in connections within and between regions, develops in growth spurts that are nearly identical to the time frame of the developmental cycles described above: These cortical connections develop in approximately four-year cycles, within which there are growth spurts in EEG coherence (a measure of cortical connectivity) that last for 6–12 months (Somsen et al., 1997). The three cycles extending from early child-hood onwards occur at approximately 1.5–5 years, 5–10 years, and 10–14 years of age, with growth spurts occurring at the ages of 6, 10, and 14 years. Along the same line, Epstein (1986) showed that spurts in EEG activity coincide with the first phase of the cycles presented here in most of the cases: they occur at the age of 2–4, 6–8, 10–12 and 14–16 years. It may thus be the case that transitions across cycles relate to the establishment of the brain networks (based on EEG coherency and power) necessary to project representational alignments of an ear-lier cycle into the more abstract networks capturing the new units that emerged from these alignments. Changes in the second half of each brain cycle relate the extending and consolidating connections within regions.

Recently, Vendetti and Bunge (2014) proposed that processing of higher order relations (e.g., analogical reasoning) engages a three-pole network that develops systematically. The first pole, located in the inferior parietal lobule (IPL), represents specific rather than general relations and it scales with the number of relations to be considered. This pole makes first-order relations available to the second pole, left rostrolateral prefrontal cortex (RLPFC). This pole abstracts second-order relations, comparing and integrating mental representations over common relations running through them. The third pole, left dorsolateral prefrontal cortex (DLPFC), pro-vides a supporting role, enabling performance monitoring, interference suppres-sion, response selection, and manipulation of items in working memory.

Further, in the 7–10 years cycle, the second to third pole network is activated but it processes both first- and second-order relations, which are not well differentiated. However, this occurs in two phases coinciding with the recycling phases described above (Wendelken et al., 2015). That is, in the 6–8 years phase, reasoning devel-opment goes with weakening of connectivity between the DLPFC and the left ventrolateral prefrontal cortex (VLPFC). Interestingly, in this period, changes in speed mediated changes in reasoning ability, suggesting that entering the cycle of rule-based reasoning is associated with differentiation between DLPFC and VLPFC, causing faster processing. In the next phase, the 9–11 years phase, the RLPF domi-nated and it was coupled with the right RLPFC. Notably, in this age phase, changes in working memory mediated changes in reasoning, suggesting that consolidation of rule based-reasoning is associated with the consolidation of the relevant brain network that is expressed in working memory expansion. In the 11–14 years phase, left and right DLPFC as well as dorsomedial PFC were more strongly engaged in processing second-order rather than first-order relations. That is, in this phase, the second pole takes its primary function in processing second-order relations. In the 15–18 years phase, left RLPFC and bilateral IPL were engaged in processing second-order relations as contrasted to first order relations. It seems that cortical reorganization within the IPL leads to greater efficiency in processing first-order relations, thereby reducing relational processing demands within RLPFC. Thus, increased communication between these brain regions over development supports their functional specialization. In this last period, the speed-reasoning and the working memory-relations disappeared, suggesting the consolidation and automation of higher-order inference (Zebec et al., 2015).

Conclusions

We summarized a theory about the architecture and development of the mind and reviewed research on brain architecture and development. The aim was to explore how these two levels of description of the human mind are interrelated. In this discussion we evaluate the evidence in reference to the three questions stated at the beginning of the section focusing on the brain.

Architectures of mind and brain. The human mind involves (i) several specialized domains of thought, and central (ii) representational, (iii) integrative, and (iv) cognizance processes. The evidence shows that there are several networks in the brain subserving each one of these mental functions. Specifically, (i) different networks carry out the core processes comprising each of the psychological domains (rooted in sensory cortices but extending into various regions depending upon the relations involved). (ii) Other networks support representational efficiency (rooted in the hippocampus and the reticular formation). (iii) Other networks take as inputs the networks above, aligning them and abstracting their commonalities (rooted in temporal, parietal and prefrontal cortices). (iv) Finally, other networks monitor, orient, select, and regulate the networks above to optimize goal-related abstractions (rooted in frontal and medial cortices) and handle differences. We emphasized brain networks rather than brain structures as bases of mental processes, because mental functions are served by overlapping, used and reused, developmentally changing, brain networks rather than single structures (Anderson, 2015).

Mental and brain processes. Meaning emerges from mapping representations onto each other so that they are compared, integrated, redefined, or re-represented into new representations. AACog is a package name standing for these processes. The brain analogue of AACog lies in the interactive and syntactic processes of the brain. Specifically, various oscillatory rhythms reflecting the activation of brain units and networks stand for "letters," "words," and "sentences" in the meaning making process. At the mental level, comparison between representations and search for similarities and differences is implemented by alignment processes. At the brain level these processes are expressed in oscillatory co-activations between the networks representing the mental entities involved.

Ideally, complete rhythm coupling would signify representational alignment. At the mental level, abstraction occurs when the commonalities between the representations aligned are identified. The brain equivalent of abstraction may be the lock of the rhythms coupled above through a rhythm of a different band, such as when several gamma oscillations are bridged by a theta oscillation. Cognizance may emerge when this new theta oscillation is made available into a broader theta-based network, thereby functioning as an autonomous token of a new mental object. The RLPFC-DLPFC-IPL circuitry is the network serving these needs (Dumontheil, 2014; Vendetti and Bunge, 2014).

Mental and brain changes. Development occurs in recurrent cycles of emergence and alignment yielding insights into the cycle's dominant mental unit and alignment process thereby letting them be re-encoded and metarepresented. This insight is embedded into phase appropriate executive-control programs that organize meaning-making and behavior. The development of this insight captures the intentional constructive aspect of intellectual development. All major theories described four levels of intellectual development with transitions at 1.5–2, 6–7, and 11–12 years (e.g., Piaget, 1970; Case, 1985), associated with increasingly abstract and better directed representations and some kind of recycling within levels standing for an early and a late phase within each cycle (Case, 1985; Fischer, 1980; Pascual-Leone, 1970; Piaget, 1970). Obviously, there must be something powerful in these transitions to have been recognized by every theorist.

The power underlying these transitions lies in the fact that brain networks change in cycles reminiscent of intellectual cycles. The development from modular core processes in infancy to integrated mental operations within each of the domains in early childhood and from these to general reasoning patterns in late childhood and adolescence correspond to successive expansions of neuronal networks such that earlier networks are integrated into the hub architecture of the networks constructed later. The crucial aspect of these expansions lies in the addition of extra connections to the parietal and the frontal hubs. Interestingly, the basic symbolic units of each cycle are discernible at both the mental and the brain level. At the mental level, they are episodic in the first cycle, representational but visually dependent in the second cycle, rule-based and language encoded in the third cycle, and principle-based and language or arbitrary symbol systems (e.g., mathematical) encoded in the fourth cycle. At the brain level, the dominant networks are located in the sensory and the motor cortices, the reticular and the parietal, the prefrontal and the frontal cortices, respectively. In other words, epigenetic mind-brain interactions transform the mind-brain system into a powerful representational machine capable to create and use complex abstract representations, in the service of different domains of knowledge.

Transitions across cycles occur when already established networks (and representations) are embedded into more complex networks. In the brain, long-distance connections between regions allow the transcription of current networks into higher-level networks that may express new relations in the input. Network expansion may be indexed by several psychological markers. Speed is a powerful marker of the initial phases of network expansion, because increases in processing speed would reflect changes in the flow of its activation until the core of the network is consolidated. After a certain point in time, the network expands to include already available instances of the lower level networks, rendering working memory capacity a better psychological marker of network expansion, because it reflects its horizontal expansion. The findings by Wendelken et al. (2015) summarized are in line with this recycling of relations between speed, working memory, and reasoning.

The fundamental components of the fronto-parietal executive control network are in place from very early in life, systematically expanding in the fashion of the four executive control programs outlined in the first section of the chapter. This is a skeletal scaffold for the development of more specialized networks, such as the various logical schemes of deductive reasoning and problem solving strategies. That is, content rich networks are built around the executive scaffold of each phase via a process of translation of the scaffold network into the domain-specific relations and constraints. Building these networks requires the activation of specific circuitry that would carry on the representation of the specific information and relations involved. Thus, speed and working memory are good indices of both developmental and individual differences because they reflect the functional state of underlying skeletal networks.

We suggested that cognizance rather than speed or working memory is the causal factor for transitions in intellectual development, because it takes mental activity as input to further mental activity. The view of consciousness as a recursive system of interactions between a central executive-selection network and other brain systems may be the system generating new mental content through its continuous rewiring. A developmental version of the global workspace model would capture how insight is built in each developmental cycle, opening the way to the next cycle. Specifically, the global workspace model implies that at any time in development, the content, resolution, and precision of awareness and cognizance depends on the state, differentiation, and synchronization of the globally synchronized networks. Thus, in each developmental phase, the awareness possible is commensurate with the network available. We showed that in each next phase increasingly more local networks are hooked onto the global workspace network and more long distance connections are added.

Obviously, there are many unanswered questions and problems to solve. For example, there is no cognitive function whose corresponding brain structures and networks are fully known. Moreover, we still do not know what is truly general and what is truly specific in both the brain and the mind. Also, we still do not know how each of the various networks carry on its own job (e.g., in terms of rhythms) and how they are integrated into a final solution behaviorally and subjectively. Finally, we know very little about how the various types of change in the brain (e.g., in myelination, electrical activity) interact with cognitive developmental changes. Therefore, the grant neurocognitive developmental theory of intelligence that would integrate brain with functional and subjective maps of mental functions into a common landscape is still far ahead of us.

References

Abu-Akel, A., & Shamay-Tsoory, S. 2011. Neuroanatomical and neurochemical bases of theory of mind. *Neuropsychologia*, 49, 2971–84.

Anderson, M. 2015. *After phrenology: Neural reuse and the interactive brain.* New York: Bradford.

Baars, B. J. 1989. *A cognitive theory of consciousness.* Cambridge, MA: Cambridge University Press.

Baddeley, A. 2012. Working memory: Theories, models, and controversies. *Annual Review of Psychology*, 63, 1–29.

Buzsaki, G., & Brendon, W. O. 2012. Brain rhythms and neural syntax: Implications for efficient coding of cognitive content and neuropsychiatric disease. *Dialogues in Clinical Neuroscience*, 14, 345–67.

Case, R. 1985. *Intellectual development: Birth to adulthood.* New York: Academic Press.

Cowey, C. M. 1996. Hippocampal sclerosis on working memory. *Memory*, 4, 19–30.

Crick, F. C., & Koch, C. 2005. What is the function of the claustrum? *Philosophical Transactions of the Royal Society: Brain and Biological Sciences*, 30, 1271–9.

Dehaene, S. 2011. *The number sense: How the mind creates mathematics.* Oxford: Oxford University Press.

Demetriou, A., & Kazi, S. 2006. Self-awareness in g (with processing efficiency and reasoning). *Intelligence*, 34, 297–317.

Demetriou, A., Mouyi, A., & Spanoudis, G. 2010. The development of mental processing. In W. F. Overton (Ed.), *Biology, cognition, and methods across the life-span. Vol. 1: Handbook of life-span development* (pp. 306–43), Editor-in-chief: R. M. Lerner. Hoboken, NJ: Wiley.

Demetriou, A., Spanoudis, G., & Shayer, M. 2014a. Inference, reconceptualization, insight, and efficiency along intellectual growth: A general theory. *Enfance*, 3, 365–96.

Demetriou, A., Spanoudis, G., Shayer, M., Mouyi, A., Kazi, S., & Platsidou, M. 2013. Cycles in speed-working memory-G relations: Towards a developmental-differential theory of the mind. *Intelligence*, 41, 34–50, doi: 10.1016/j.intell.2012.10.010.

Demetriou, A., Spanoudis, G., Shayer, M., Van Der Ven, S., Brydges, C. R., Kroesbergen, E., Podjarny, G., & Swanson, H. L. 2014b. Relations between speed, working memory, and intelligence from preschool to adulthood: Structural equation modeling of 15 studies. *Intelligence*, 46, 107–21.

Dumontheil, I. 2014. Development of abstract thinking during childhood and adolescence: The role of rostrolateral prefrontal cortex. *Developmental Cognitive Neuroscience*, 10, 57–76.

Edelman, G. M., & Tononi, G. A. 2000. *Universe of consciousness.* New York: Basic Books.

Fischer, K. W. 1980. A theory of cognitive development: The control and construction of hierarchies of skills. *Psychological Review*, 87, 477–531.

Fodor, J. A. 1975. The language of thought. *Hassocks: Harvester Press.*

Fonlupt, P. 2003. Perception and judgement of physical causality involve different brain structures. *Cognitive Brain Research*, 17, 248–54.

Franson, P., Aden, U., Blenow, M., & Lagercrantz, H. 2011. The functional architecture of the infant brain as revealed by resting-state fMRI. *Cerebral Cortex*, 21, 145–54.

Friedman, H. R., & Goldman-Rakic, P. S. 1988. Activation of the hippocampus and dentate gyrus by working memory: A two-deoxyglucos study of behaving rhesus monkeys. *The Journal of Neuroscience*, 8, 4693–706.

Galaburda, A. M. 2002. The neuroanatomy of categories. In A. M. Galaburda, S. M. Kosslyn, & Y. Christen (Eds.), *The languages of the brain* (pp. 23–42). Cambridge, MA: Harvard University Press.

Goel, V., Buchel, C., Frith, C., & Dolan, R. J. 2000. Dissociation of mechanisms underlying syllogistic reasoning. *NeuroImage*, 12, 504–14.

Gruber, O., & von Cramon, Y. 2003. The functional neuroanatomy of human working memory revisited. Evidence from 3-T fMRI studies using classical domain-specific interference tasks. *Neuroimage*, 19(3), 797–809.

Gruber, O., & Goschke, T. 2004. Executive control emerging from dynamic interactions between brain systems mediating language, working memory, and attentional processes. *Acta Psychologica*, 115, 105–21.

Harris, I. M., Egan, G. F., Sonkkila, C., Tochon-Danguy, H. J., Paxinos, G., & Watson, D. G. 2000. Selective right parietal lobe activation during mental rotation. *Brain*, 123, 65–73.

Hoff, G. E. A-J., Van Den Heuvel, M. P., Benders, M. J. N. L., Kersbergen, K. J., & De Vries, L. S. 2013. On development of functional brain connectivity in the young brain. *Frontiers in Human Neuroscience*, 6, 650; doi:10.3389/fnhum.2013.00650.

Hwang, K., Hallquist, M. N., & Luna, B. 2012. The development of hub architecture in the human functional brain network. *Cerebral Cortex*, 23(10), 2380–93; doi:10.1093/cercor/bhs227.

Jensen, A. R. 1998. *The g factor: The science of mental ability*. Westport, CT: Praeger.

Jung, R. E., & Haier, R. J. 2007. The Parieto-Frontal Integration Theory (P-FIT) of intelligence: Converging neuroimaging evidence. *Behavioral and Brain Sciences*, 30, 135–87.

Kaldy, Z., & Leslie, A. M. 2005. A memory span of one? Object identification in 6.5-month-old infants. *Cognition*, 97, 153–77.

Kanwisher, N., Mcdermott, J., & Chun, M. M. 1997. The Fusiform Face Area: A module in human extrastriate cortex specialized for face perception. *The Journal of Neuroscience*, 17, 4302–11.

Klingberg, T. 1998. Concurrent performance of two working memory tasks: Potential mechanisms of interference. *Cerebral Cortex*, 8, 593–601.

Lisman, J. 2005. The theta/gamma discrete phase code occurring during the hippocampal phase precession may be a more general brain coding scheme. *Hippocampus*, 15, 913–22.

Menon, V., & Uddin, L. Q. 2010. Saliency, switching, attention, and control: A network model of insula function. *Brain Structure and Function*, 214, 655–67.

Morra, S., Alp, E., Panesi, S., & Viterbori, P. 2014. Working memory and its impact on cognitive development in very young children. Presented at the 7th European Working Memory Symposium Edinburgh, September 2–4.

Pascual-Leone, J. 1970. A mathematical model for the transition rule in Piaget's developmental stages. *Acta Psychologica*, 63, 301–45.

Petersen S., & Posner, M. I. 2012. The attention system of the human brain: 20 years after. *Annual Review of Neuroscience*, 35, 73–89.

Piaget, J. 1970. Piaget's theory. In P. H. Mussen (Ed.), *Carmichael's handbook of child development* (pp. 703–32). New York: Wiley.

Pickering, S. J. 2001. The development of visuo-spatial working memory. *Memory*, 9, 423–32.

Raganath, C., & D' Esposito, M. 2005. Directing the mind's eye. Prefrontal, inferior, and medial temporal mechanisms for visual working memory. *Current Opinion in Neurobiology*, 15, 175–82.

Repovš, G., & Baddeley, A. 2006. The multi-component model of working memory: Explorations in experimental cognitive psychology. *Neuroscience*, 139, 5–21.

Sauseng, P., Griesmayer, B., Freunberger, R., & Klimesch, W. 2010. Control mechanisms in working memory: A possible function of EEG theta oscillations. *Neuroscience and Biobehavioral Reviews*, 34, 1015–22.

Sergent, C., & Naccache, L. 2012. Imaging neural signatures of consciousness: "What," "When," "Where," and "How" does it work? *Archives Italiennes de Biologie*, 150, 91–106.

Siegal, M., & Varley, R. 2002. Neuronal systems involved in "theory of mind." *Nature Reviews*, 3, 463–71.

Somsen, R. J. M., Van't Klooster, B. J., Van Der Molen, M. W., Van Leeuwen, H. M. P., & Licht, R. 1997. Growth spurts in brain maturation during middle childhood as indexed by EEG power spectra. *Biological Psychology*, 44, 187–209.

Spanoudis, G., Demetriou, A., Kazi, S., Giorgala, K., & Zenonos, V. 2015. Embedding cognizance in intellectual development. *Journal of Experimental Child Psychology*, 132, 32–50.

Squire, L. R. 1992. Memory and the hippocampus: A synthesis from rats, monkeys, and humans. *Psychological Review*, 92, 195–231.

Thatcher, R. W. 1992. Cyclic cortical reorganization during early childhood. *Brain and Cognition*, 20, 24–50.

Thatcher, R. W. 1994. Cyclic cortical reorganization: Origins of human cognitive development. In G. Dawson and K. Fischer (Eds.), *Human Behavior and the Developing Brain* (pp. 232–66). New York: Guilford.

Vendetti, M. S., & Bunge, S. A. 2014. Evolutionary and developmental changes in the lateral frontoparietal network: A little goes a long way for higher-level cognition. *Neuron*, 84, 906–17.

Vergauwe, E., Hartstra, E., Barrouillet, P., & Bras, M. 2015. Domain-general involvement of the posterior frontolateral cortex in time-based resource-sharing in working memory: An fMRI study. *NeuroImage*, 115, 104–6; doi.org/10.1016/j.neuroimage.2015.04.059.

Wendelken, C., Ferrer, E., Whitaker, K. J., & Bunge, S. A. 2015. Fronto-parietal network reconfiguration supports the development of reasoning ability. *Cerebral Cortex*, 1–13, doi: 10.1093/cercor/bhv050.

Zebec, M., Demetriou, A., & Topic, M. 2015. Changing expressions of general intelligence in development: A two-wave longitudinal study from 7 to 18 years of age. *Intelligence*, 49, 94–109.

3 Cognitive capacities of the infant mind

A neuroimaging perspective[1]

Mohinish Shukla and Vivian Ciaramitaro

Introduction

The average layperson, if asked whether humans are conscious, would probably reply in the affirmative. But if, instead of "humans" we asked if the "human infant" is conscious, it's not clear what the average lay person would say. As recently as 1890, William James proposed that the sensory world of the infant was an undifferentiated mess, which he colorfully described as a "blooming, buzzing confusion" (James, 1890/1981). Indeed, in the behaviorist tradition, and its modern incarnations, the infant is proposed to have very little actual mental content, and that content is slowly acquired from the environment using (sophisticated) general learning mechanisms. By mental content, we mean, among other things, constructs, such as concepts, thoughts, beliefs, or desires, that are intentional, and to some extent represent knowledge about the world. If the mind of an infant is indeed a blooming, buzzing confusion, then it's not clear what substantial stuff they could be conscious *of*. For example, it wasn't until the 1980s that infants were given anything more than token anesthetics when undergoing a surgical intervention. The assumption was that infants could not be conscious of pain. It took empirical studies documenting the negative consequences of operating on infants without anesthetics to make pediatric anesthetic use commonplace (Anand and Hickey, 1987).

Clearly, the idea that infants are conscious has profound impacts in how we treat them. While the use of pediatric anesthesia is an extreme example, it is still important to consider how we see infants in terms of their cognitive capacities. Take the case of the language we speak. In the 1964 movie classic *My Fair Lady*, Professor Henry Higgins suggests that it's the flower-girl Eliza's Cockney accent that marks her low social class and that if the Colonel Pickering were to speak as she did, why, he might be selling flowers too. Indeed, social psychologists have documented that the language one speaks has a profound influence on how one is viewed by others (e.g., Gluszek and Dovidio, 2010). How, then, should we view infants, who cross-culturally do not appear to have any language at all? Can we safely assume that they have no conception of language at all (as is common in the empiricist tradition), or should we assume that they have a broad and general understanding of language, that is subsequently narrowed to the language in their environment? The same question can be asked of other cognitive capacities that we believe human adults possess, including knowledge of objects and events, of numbers, rationality, and even basic social skills.

In this chapter, we will examine studies that have looked at cognitive capacities in infants in several such domains, with a special focus on converging evidence provided by behavioral and neuroimaging studies. The neuroimaging studies we review utilize electrophysiology techniques (electroencephalography, EEG, and event-related potentials, ERPs; see Srinivasan and Nunez, 2012, for an overview), and hemodynamic measurements (including functional magnetic resonance imaging, fMRI, and functional near-infrared spectroscopy, fNIRS; see Aslin et al., 2015, for a recent review of the use of hemodynamic measures in studying infant cognition). Our coverage of these studies is not intended to be complete; we have instead selected studies that parallel and reinforce each other. We cannot say with any certainty whether or not infants are conscious of objects, social relations, etc., but we can at least say that they are capable of certain relevant computations in these domains. We will primarily highlight the processing of objects in the physical domain, and of faces and language in the social domain.

Object representation and individuation

A fundamental distinction in how we see the world, is as *objects* and *actions*. That is, our perceptual systems encode some algorithms for parsing and grouping the perceptual world into discrete entities, and then assign these entities properties that can be static (like their shape or color), or dynamic (like hopping or slithering). Several lines of evidence suggest that infants start out with impoverished representations of objects, which they subsequently refine through experience (e.g., Baillargeon et al., 2012), although some classes of stimuli, like human faces, appear to be have more detailed initial representations (see next section).

Behavioral studies have been largely concerned with understanding early cognitive capacities and constraints on such *object individuation*. However, methodologically speaking, it is difficult to understand how an infant is parsing a scene that she is watching, so researchers have primarily relied on paradigms that require infants to parse a scene into relevant object representations *and* hold these representations in memory, for subsequent recall. Using such encoding-and-recall behavioral paradigms, researchers have found evidence for *object permanence*—the ability to maintain a memory trace of a previously seen object, in infants as young as 2–3 months of age (e.g., Aguiar and Baillargeon, 1999). In a typical study, infants are exposed to small objects that appear and disappear from either side of an opaque screen. Researchers vary properties of the appearing and disappearing objects to probe the nature of what features of an object infants encode. For example, a red ball might emerge from the left edge of an opaque occluder, and then turn back and disappear behind it. Subsequently, if a red cube similarly emerges and disappears from the right edge of the opaque occluder, infants around 7 months of age are surprised (look longer) if the opaque occluder drops to reveal only the red cube, suggesting that they were presumably expecting two objects. A change in color alone (e.g., using a green ball instead of the red cube from the right edge) does not lead to such a surprise reaction, suggesting that infants' early object representations preferentially encode shape over color. Recent work however suggests that the demands for parsing and those for encoding in memory might be in conflict, specially for infants with their limited resources, such that infants might need to prioritize encoding objects versus encoding the features of the objects (Kibbe and Leslie, 2013).

Turning to neuroimaging, EEG measures in 7- to 12-month-olds have been shown to correlate with better behavioral performance on object permanence tasks (Bell and Fox, 1992). In this study, resistance to increasing delays between the disappearance of the final object and the lowering of the opaque screen (indicating a stronger memory trace of the object) were accompanied by stronger neuronal activity or bigger frontal EEG power. fNIRS has also been used to study object permanence in infants (Baird et al., 2002). In a series of linked studies, Wilcox and colleagues have used fNIRS to investigate hemodynamic responses in 7-month-old infants. Using paradigms similar to the ones described above, these authors found that infants react to changes in shape and texture, but not in color; and that these behavioral findings are reflected in increased hemodynamic responses for shape and texture changes compared to color changes, in fNIRS channels placed over the temporal cortex (see Wilcox and Biondi, 2015, for a recent review).

Related to infants' early capacity for object individuation, is their capacity to compute *how many* objects are present in an array. A wealth of experimental and theoretical work has revealed that there are two core systems for numerical thinking—one system (Core 1) for representing the approximate magnitude of large collections of objects, and a second system (Core 2) for representing the precise cardinality of sets of up to three objects (e.g., Feigenson et al., 2004).

Studies have carefully controlled that behavioral results implicating a representation of numerosity do not arise from confounds introduced by changes in low-level stimulus features that can co-vary with the number of objects displayed, such as size, surface area, or density (Lipton and Spelke, 2003; Xu and Spelke, 2000). Furthermore, studies have shown that numerical discriminations are independent of the sensory modality of the input; infants can discriminate quantities presented by either visual or auditory objects (Izard et al., 2009; Jordan and Brannon, 2006; Lipton and Spelke, 2003).

Recently, non-invasive neuroimaging methodologies have been used to investigate the neural correlates of the number sense, revealing many of the same underlying brain networks for numerosity in adulthood and in the developing brain. In adults, the intraparietal sulcus in the posterior parietal region has been shown to represent approximate numerical information, irrespective of the sensory modality of the input, or the use of symbolic or non-symbolic representations (for review see Dehaene et al., 2004). Furthermore, small numerosities were encoded closer to the midline, while larger quantities were encoded more laterally, suggesting that the behavioral difference between the two core systems is accompanied by a corresponding anatomical difference in localization.

Just as in adults, infants also show activity in parietal cortex when performing non-verbal numerical tasks. A recent fNIRS study by Hyde and colleagues finds that the parietal cortex is active in 6-month-old infants, an age before language development and before experience with symbolic numerical representations (Hyde et al., 2010). Turning to ERP studies in infants, Hyde and Spelke (2011) found that 6- to 7.5-month-olds showed distinct ERP components sensitive to precise, small numbers versus approximate, large numbers. ERP evidence also suggests selective processing in parietal cortex for numerosity as opposed to other low-level sensory features (Izard et al., 2008). Thus, specialization for number in the infant brain can be seen independently of any formal mathematical experience or education or even before language.

Social cognition

Turning away from the material world of objects and their numerosities, let us now consider the human as a social animal. As such, we would expect human infants to display early social capacities. That is, in addition to bringing some primitive computations about their physical world, we would expect infants to also bring some primitive computations relevant to their social world. Indeed, newborn infants show preferences for socially relevant stimuli, in particular faces, described below, and speech and language, described in the next section.

A face is an important, socially relevant stimulus category not just because it signals the identity of a person, but also because it conveys information about mental states—for example, those expressed through emotions. The standard view of the development of face recognition had been that the ability to discriminate different faces was not present until 10 years of age and that such an ability was driven primarily by experience (Carey and Diamond, 1977; Carey et al., 1980). Classic studies in newborns have long since discredited this view, providing evidence that even newborns have the machinery for face recognition. In one such study, Goren et al. (1975) found that newborn infants preferentially oriented towards face-like stimuli with primitive "eye" and "mouth"-like features. Infants did not prefer such stimuli when they were presented upside-down, suggesting a particular sensitivity to specific feature arrangements that might be indicative of faces in their visual input. Subsequent experiments have refined this notion of an early face-like template, to include features like the contrast polarity of the eyes (black pupils on a white background; Farroni et al., 2005).

In adults, evidence for a specialized mechanism for face processing comes both from studies of prosopagnosia—a neuropsychological condition with a specific deficit in the visual processing of faces (e.g., McNeil and Warrington, 2013)—and imaging studies that have pinpointed a specific brain region, the fusiform face area (FFA: Kanwisher et al., 1997). Unfortunately, the FFA is inaccessible using fNIRS owing to its distance from the scalp, and there have not been many studies using fMRI with very young children (but see Cantlon et al., 2011). Researchers have used fNIRS to examine additional neural areas that might be involved in processing face stimuli in 5- to 8-month-old infants. These studies examine activations to faces compared to control stimuli, including inverted faces or pictures of vegetables (Otsuka et al., 2007), and typically find that faces preferentially engage lateral (temporal) cortices. Although Otsuka et al. (2007) found a right-hemisphere (RH) advantage for faces, subsequent studies found engagement of bilateral temporal areas, and additionally showed that these areas show higher activations for blocks of several different faces, compared to blocks with a single, repeated face, and also show size and viewpoint invariance (Kobayashi et al., 2011, 2012).

ERP studies find that infants as young as 3 months of age show enhanced amplitude of a particular component, the N290, for upright relative to inverted faces; such an inversion effect is not found for upright versus inverted cars (Peykarjou and Hoehl, 2013). Further, different ERP components have been found to be active in 7-month-old infants when viewing happy versus fearful or angry faces (Nelson and De Haan, 1996; Leppänen et al., 2007). Jessen and Grossmann (2015) relied on previous behavioral and ERP studies in adults and ERP studies in 5- and 7-month-old infants (Kouider et al., 2013) to define supra- and subliminal face stimuli, and found that neuronal responses in 7-month-old infants can not only discriminate between happy versus fearful facial expressions but that these neuronal responses also depend on whether the information about the face is being processing subliminally (subconsciously) or supraliminally (consciously).

Put together, the imaging studies provide converging evidence that in their first year of life infants are particularly tuned to various aspects of a critical, socially relevant visual stimulus, a human face. In the next section we will turn to language, probably the most important human feature, and of critical importance in all social relations.

Language

Language is considered to be a defining feature of our species (Maynard Smith and Szathmáry, 1995; Szathmáry, 2015). Language plays a particularly crucial role in the study of the mind and consciousness, as it allows the communication of mental content, including knowledge and beliefs. As noted in the introduction, a lack of language could potentially be interpreted as a lack of mental content, which makes the study of language capacities in infants especially important.

The preponderance of studies in infant language research primarily document changes in behavioral and neurophysiological responses to speech-like and non-speech-like stimuli, in order to construct theories for the development of various linguistic competencies and their brain bases (e.g., Minagawa-Kawai et al., 2011). Indeed, behavioral studies have revealed that speech is perceived as a special kind of input by young infants. Infants show a preference for speech over other sounds, including backward speech, and speech engages special processing algorithms, including categorization of sets of visual objects at 3 to 4 months of age (e.g., Ferry et al., 2010) or cueing referential communication at 6 months (Senju and Csibra, 2008).

Infant imaging studies have revealed that speech shows asymmetric patterns of activity across the two brain hemispheres, indicating cortical specialization for speech stimuli, even at birth (Peña et al., 2003; Dehaene-Lambertz et al., 2002). In particular, compared to non-speech, speech preferentially activates the left hemisphere. In one such study, (Peña et al., 2003) studied neonates using fNIRS to examine patterns of cortical activations to normal, *Forward* speech, compared to *Backward* speech (in which the same recorded speech samples are played backwards); and to *Silence*, when no stimulus is presented. These authors showed that the neonate brain already showed cortical specialization: while channels over the right temporal cortex showed similar activations to both *Forward* and *Backward* speech, channels over the left temporal cortex showed preferential activation for *Forward* over *Backward* speech.

Similar studies with fMRI have not only found left-lateralized responses specifically to speech, but have also identified other, frontal areas that are asymmetrically active for speech as compared to non-speech in older infants (2- to 3-month-olds in Dehaene-Lambertz et al., 2002; 21-month-olds in Redcay et al., 2008).

Likewise, studies using EEG have found greater left hemisphere activity when infants listen to speech and greater right hemisphere activity when infants listen to non-speech sounds, such as noises and piano chords in one-week to 10-month-old infants (Molfese et al., 1975; also see Molfese and Molfese, 1979, 1980, 1985). Furthermore, the left-lateralization of responses is present not only for spoken, but also for signed languages (Petitto et al., 2000). Importantly, ERP measures to speech in newborns are predictive of later language proficiency (Molfese and Molfese, 1985, 1997).

Not all aspects of speech perception are left lateralized. In particular, it has been proposed that the left hemisphere is primarily responsible for computations over shorter time-scales, while the right hemisphere is responsible for computations over longer time-scales (see reviews in Poeppel, 2003; Hickok and Poeppel, 2007). This division of labor in terms of the temporal aspect of speech corresponds with the proposed division of the phonological organization of speech sounds at the segmental (e.g., consonants and vowels) level and the suprasegmental, prosodic level, including changes in pitch, intensity, and duration. For example, a sentence like "It is raining" can be spoken as a statement or a question—in both cases, the segmental level is constant, while the prosody varies. Neurophysiological studies with infants have shown preferential activation for prosody both at the level of the sentence (e.g., Homae et al., 2006, with Japanese 3-month-olds) and at the level of the word (e.g., Weber et al., 2004, with German 5-month-olds).

Language is much more than speech. Speech acts are but the linear externalizations of abstract, hierarchically organized grammatical structures. Although most imaging studies have investigated the processing of speech, few have addressed the question—can the mind/brain of the young infant extract abstract structure from the speech input? Starting with the now-classic study by Marcus and colleagues (Marcus et al., 1999), it has been shown that infants can extract simple structures from short exposures to carefully controlled artificial speech stimuli. For example, when trained with trisyllabic sequences such as "de-li-li," 7-month-old infants subsequently discriminated novel sequences that followed the A-B-B pattern (such as "wo-fe-fe") from ones that did not (such as "wo-fe-wo"). Recently, the impressive power of 7-month-olds' abstract processing capacity was revealed in a study that asked if infants could represent an A-B-B pattern if the elements of the pattern were not just single syllables, as in the original Marcus study, but were themselves sequences that followed one rule or another (Kovács and Endress, 2014). For example, an A element could be "wo-fe-fe" or "de-li-li," while a B element could be "wo-fe-ga" or "de-li-me." Remarkably, 7-month-olds were able to learn these rules-over-rules, suggesting that infants possess some ability to represent their perceptual input as hierarchically organized structures.

Imaging studies have examined neural responses that accompany the extraction of structure from organized, artificial speech stimuli. Using fNIRS, Gervain et al. (2008) found that trisyllable blocks (that is, a series of trisyllables separated by silence) that had an A-B-B structure showed greater activation in left cortical areas compared to random syllable triplets (an A-B-C structure). Further, the left-lateralized cortical response over frontal areas showed a progressive enhancement over the experimental period, suggesting that frontal areas might have been responding to increased familiarity with the structure, indicating not just analysis, but possibly also learning (see also Gervain et al., 2012; Wagner et al., 2011).

Taken together, the behavioral and neuroimaging data provide converging evidence that the newborn baby's brain is prepared to rapidly identify, analyze, and learn from socially important stimuli in its perceptual input, whether this be faces (previous section) or language.

Conclusions

Imagine a boulder hurtling down a mountain path that forks just ahead of the boulder.[2] When we see the boulder take the path on its left instead of the one on its right, we don't infer that the boulder made a "choice," but that small variations in the physical properties of the path just before the fork led to the boulder following one path over the other, in a purely deterministic fashion. Now imagine a runner sprinting down the path and also taking the path on her left. In this case, we are likely to attribute the selection of the path to some internal mental state like a belief or a desire. That is, purely behavioral observations are not sufficient to infer mental states. Similarly, researchers who study consciousness are acutely aware of the seemingly intractable philosophical problem of trying to study something as intangible as mental states through video recordings or measurements of voltage distributions across the scalp (e.g., Kouider et al., 2013).

Instead of addressing this complex question, we can ask—is the mental life and the accompanying brain physiology of infants significantly different from that of adults? In this chapter we have tried to summarize behavioral and imaging evidence that suggests that, right from birth, infants display evidence for possessing sophisticated cognitive capacities in several cognitive domains. Moreover, these capacities are equivalent to similar capacities in adults, either in their behavioral outcomes, in their physiology, or both. Given these substantial similarities, and given that the baby human will grow into an adult human, it is reasonable to ask, is an infant merely an adult without the years of experience? The results summarized here suggest that this might indeed be the case. That is, the similarities between infants and adults ought to compel us to recognize that the mental life of an infant might be substantially similar to that of an adult, however the mental life of an adult is constructed. For example, if a linguistic theory of the adult language faculty requires constructs such as nouns or verbs, then there is no a priori reason to suspect that these same constructs are not part of the mental life of infants. In his book *The Prism of Grammar*, Tom Roeper gathers together a wealth of evidence to show that the earliest utterances of infants reflect deep, abstract structures, and creative processes that form the basis of human thought (Roeper, 2007). We hope that our survey of the cognitive abilities of infants, as assessed through behavioral of physiological methods, provide further impetus to the idea that our infants are humans much like us, and thereby afford them the same dignity that is extended to all other members of our species.

Notes

1 Substantial parts of this chapter are reproduced with permission from Aslin, Shukla, and Emberson (2015). Hemodynamic correlates of cognition in human infants. *Annu Rev Psychol*, 66, 349–379.
2 This analogy is loosely borrowed from Roeper (2007), Ch. 12.

References

Aguiar, A., & Baillargeon, R. 1999. 2.5-month-old infants' reasoning about when objects should and should not be occluded. *Cogn Psychol*, 39(2), 116–57. doi: 10.1006/cogp.1999.0717.

Anand, K., & Hickey, P. 1987. Pain and its effects in the human neonate and fetus. *New Engl J Med*, 317(21), 1321–9.

Aslin, R. N., Shukla, M., & Emberson, L. L. 2015. Hemodynamic correlates of cognition in human infants. *Annu Rev Psychol*, 66, 349–79.

Baillargeon, R., Stavans, M., Wu, D., Gertner, Y., Setoh, P., Kittredge, A., & Bernard, A. 2012. Object individuation and physical reasoning in infancy: An integrative account. *Lang Learn Dev*, 8(1), 4–46.

Baird, A. A., Kagan, J., Gaudette, T., Walz, K. A., Hershlag, N., & Boas, D. A. 2002. Frontal lobe activation during object permanence: Data from near-infrared spectroscopy. *Neuroimage*, 16(4), 1120–5.

Bell, M., & Fox, N. 1992. The relations between frontal brain electrical activity and cognitive development during infancy. *Child Dev*, 63, 1142–63.

Cantlon, J., Pinel, P., Dehaene, S., & Pelphrey, K. 2011. Cortical representations of symbols, objects, and faces are pruned back during early childhood. *Cereb Cortex*, 21(1), 191–9. doi: 10.1093/cercor/bhq078.

Carey, S., & Diamond, R. 1977. From piecemeal to configurational representation of faces. *Science*, 195(4275), 312–14.

Carey, S., Diamond, R., & Woods, B. 1980. Development of face recognition: A maturational component? *Dev Psychol*, 16(4), 257.

Dehaene, S., Molko, N., Cohen, L., & Wilson, A. J. 2004. Arithmetic and the brain. *Curr Opin Neurobiol*, 14(2), 218–24.

Dehaene-Lambertz, G., Dehaene, S., & Hertz-Pannier, L. 2002. Functional neuro-imaging of speech perception in infants. *Science*, 298(5600), 2013–15.

Farroni, T., Johnson, M. H., Menon, E., Zulian, L., Faraguna, D., & Csibra, G. 2005. Newborns' preference for face-relevant stimuli: Effects of contrast polarity. *Proc Natl Acad Sci U S A*, 102(47), 17245–50. doi: 10.1073/pnas.0502205102.

Feigenson, L., Dehaene, S., & Spelke, E. 2004. Core systems of number. *Trends Cogn Sci*, 8(7), 307–14.

Ferry, A. L., Hespos, S. J., & Waxman, S. R. 2010. Categorization in 3-and 4-month-old infants: An advantage of words over tones. *Child Dev*, 81(2), 472–9.

Gervain, J., Berent, I., & Werker, J. F. 2012. Binding at birth: The newborn brain detects identity relations and sequential position in speech. *J Cogn Neurosci*, 24(3), 564–74.

Gervain, J., Macagno, F., Cogoi, S., Peña, M., & Mehler, J. 2008. The neonate brain detects speech structure. *Proc Natl Acad Sci U S A*, 105(37), 14222–7. doi: 10.1073/pnas.0806530105.

Gluszek, A., & Dovidio, J. 2010. The way they speak: A social psychological perspective on the stigma of nonnative accents in communication. *Pers Soc Psychol Rev*, 14(2), 214–37.

Goren, C. C., Sarty, M., & Wu, P. Y. 1975. Visual following and pattern discrimination of face-like stimuli by newborn infants. *Pediatrics*, 56(4), 544–9.

Hickok, G., & Poeppel, D. 2007. The cortical organization of speech processing. *Nat Rev Neurosci*, 8, 393–402.

Homae, F., Watanabe, H., Nakano, T., Asakawa, K., & Taga, G. 2006. The right hemisphere of sleeping infant perceives sentential prosody. *Neurosci Res*, 54(4), 276–80. doi: 10.1016/j.neures.2005.12.006.

Hyde, D. C., Boas, D. A., Blair, C., & Carey, S. 2010. Near-infrared spectroscopy shows right parietal specialization for number in pre-verbal infants. *Neuroimage*, 53(2), 647–52. doi: 10.1016/j.neuroimage.2010.06.030.

Hyde, D. C., & Spelke, E. S. 2011. Neural signatures of number processing in human infants: Evidence for two core systems underlying numerical cognition. *Developmental Sci*, 14(2), 360–71.

Izard, V., Dehaene-Lambertz, G., & Dehaene, S. 2008. Distinct cerebral pathways for object identity and number in human infants. *PLoS Biol*, 6(2), e11.

Izard, V., Sann, C., Spelke, E. S., & Streri, A. 2009. Newborn infants perceive abstract numbers. *P Natl Acad Sci*, 106(25), 10382–5.

James, W. 1890/1981. *The principles of psychology*. Cambridge, MA: Harvard University Press.

Jessen, S., & Grossmann, T. 2015. Neural signatures of conscious and unconscious emotional face processing in human infants. *Cortex*, 64, 260–70.

Jordan, K. E., & Brannon, E. M. 2006. The multisensory representation of number in infancy. *P Natl A Sci USA*, 103(9), 3486–9.

Kanwisher, N., Mcdermott, J., & Chun, M. M. 1997. The fusiform face area: A module in human extrastriate cortex specialized for face perception. *J Neurosci*, 17(11), 4302–11.

Kibbe, M., & Leslie, A. 2013. What's the object of object working memory in infancy? Unraveling "what" and "how many." *Cognitive Psychol*, 66(4), 380–404.

Kobayashi, M., Otsuka, Y., Kanazawa, S., Yamaguchi, M. K., & Kakigi, R. 2012. Size-invariant representation of face in infant brain: An fNIRS-adaptation study. *Neuroreport*, 23(17), 984–8. doi: 10.1097/WNR.0b013e32835a4b86.

Kobayashi, M., Otsuka, Y., Nakato, E., Kanazawa, S., Yamaguchi, M. K., & Kakigi, R. 2011. Do infants represent the face in a viewpoint-invariant manner? Neural adaptation study as measured by near-infrared spectroscopy. *Front Hum Neurosci,* 5, 153. doi: 10.3389/fnhum.2011.00153.

Kouider, S., Stahlhut, C., Gelskov, S. V., Barbosa, L. S., Dutat, M., De Gardelle, V., & Dehaene-Lambertz, G. 2013. A neural marker of perceptual consciousness in infants. *Science,* 340(6130), 376–80.

Kovács, Á. M., & Endress, A. D. 2014. Hierarchical processing in seven-month-old infants. *Infancy,* 19(4), 409–25.

Leppänen, J. M., Moulson, M. C., Vogel-Farley, V. K., & Nelson, C. A. 2007. An ERP study of emotional face processing in the adult and infant brain. *Child Dev,* 78(1), 232–45.

Lipton, J. S., & Spelke, E. S. 2003. Origins of number sense large-number discrimination in human infants. *Psychol Sci,* 14(5), 396–401.

Marcus, G. F., Vijayan, S., Bandi Rao, S., & Vishton, P. M. 1999. Rule learning by seven-month-old infants. *Science,* 283(5398), 77–80.

Maynard Smith, J., & Szathmáry, E. 1995. The major evolutionary transitions. *Nature,* 374, 227–32.

Mcneil, J. E., & Warrington, E. K. 2013. Prosopagnosia: A face-specific disorder. *Q J Exp Psychol A,* 46(1), 1–10.

Minagawa-Kawai, Y., Van Der Lely, H., Ramus, F., Sato, Y., Mazuka, R., & Dupoux, E. 2011. Optical brain imaging reveals general auditory and language-specific processing in early infant development. *Cereb Cortex,* 21(2), 254–61. doi: 10.1093/cercor/bhq082.

Molfese, D. L., Freeman, R. B., & Palermo, D. S. 1975. The ontogeny of brain lateralization for speech and non-speech stimuli. *Brain Lang,* 2, 356–68.

Molfese, D. L., & Molfese, V. J. 1979. Hemisphere and stimulus differences as reflected in the cortical responses of newborn infants to speech stimuli. *Dev Psychol,* 15(5), 505.

Molfese, D. L., & Molfese, V. J. 1980. Cortical response of preterm infants to phonetic and nonphonetic speech stimuli. *Dev Psychol,* 16(6), 574.

Molfese, D. L., & Molfese, V. J. 1985. Electrophysiological indices of auditory discrimination in newborn infants: The bases for predicting later language development? *Infant Behav Dev,* 8(2), 197–211.

Molfese, D. L., & Molfese, V. J. 1997. Discrimination of language skills at five years of age using event-related potentials recorded at birth. *Dev Neuropsychol,* 13(2), 135–56.

Nelson, C. A., & De Haan, M. 1996. Neural correlates of infants' visual responsiveness to facial expressions of emotion. *Dev Psychobiol,* 29(7), 577–95.

Otsuka, Y., Nakato, E., Kanazawa, S., Yamaguchi, M. K., Watanabe, S., & Kakigi, R. 2007. Neural activation to upright and inverted faces in infants measured by near infrared spectroscopy. *Neuroimage,* 34(1), 399–406. doi: 10.1016/j.neuroimage.2006.08.013.

Peña, M., Maki, A., Kovacić, D., Dehaene-Lambertz, G., Koizumi, H., Bouquet, F., & Mehler, J. 2003. Sounds and silence: An optical topography study of language recognition at birth. *Proc Natl Acad Sci U S A,* 100(20), 11702–5. doi: 10.1073/pnas.1934290100.

Petitto, L. A., Zatorre, R. J., Gauna, K., Nikelski, E., Dostie, D., & Evans, A. C. 2000. Speech-like cerebral activity in profoundly deaf people processing signed languages: Implications for the neural basis of human language. *P Natl Acad Sci,* 97(25), 13961–6.

Peykarjou, S., & Hoehl, S. 2013. Three-month-olds' brain responses to upright and inverted faces and cars. *Dev Neuropsychol*, 38(4), 272–80.

Poeppel, D. 2003. The analysis of speech in different temporal integration windows: Cerebral lateralization as "asymmetric sampling in time." *Speech Commun*, 41(1), 245–55.

Redcay, E., Haist, F., & Courchesne, E. 2008. Functional neuroimaging of speech perception during a pivotal period in language acquisition. *Developmental Sci*, 11(2), 237–52.

Roeper, T. 2007. *The prism of grammar: How child language illuminates humanism.* The MIT Press.

Senju, A., & Csibra, G. 2008. Gaze following in human infants depends on communicative signals. *Curr Biol*, 18(9), 668–71.

Srinivasan, R., & Nunez, P. L. 2012. Electroencephalography. In V. S. Ramachandran (Ed.), *Encyclopedia of human behavior*, pp. 15–23. Burlington, MA: Academic Press. 2nd ed.

Szathmáry, E. 2015. Toward major evolutionary transitions theory 2.0. *P Natl Acad Sci*, 112(33), 10104–11.

Wagner, J. B., Fox, S. E., Tager-Flusberg, H., & Nelson, C. A. 2011. Neural processing of repetition and non-repetition grammars in 7- and 9-month-old infants. *Front Psychol*, 2, 168. doi: 10.3389/fpsyg.2011.00168.

Weber, C., Hahne, A., Friedrich, M., & Friederici, A. D. 2004. Discrimination of word stress in early infant perception: Electrophysiological evidence. *Cognitive Brain Res*, 18(2), 149–61.

Wilcox, T., & Biondi, M. 2015. Object processing in the infant: Lessons from neuroscience. *Trends Cogn Sci*. 19(7), 406–13.

Xu, F., & Spelke, E. S. 2000. Large number discrimination in 6-month-old infants. *Cognition*, 74(1), B1–11.

4 Neural infantese

Detecting pain and suffering in preverbal infants by means of neuro-technological communication

Karl Sallin

Introduction

Pain is an unpleasant sensory and emotional experience associated with actual or potential tissue damage (Merskey and Bogduk, 1994). Suffering refers to the emotional feelings related to long-term effects of having pain (Price, 2000). The traditional method of evaluating pain is by communicating directly, physician and patient. The narrative method allows for an individual evaluation of the physical, affective, cognitive, and social consequences of pain and suffering. However, in many clinical situations, direct communication is impeded. Also, since pain is a subjective phenomenon, it resists objective quantification. Alternative methods of pain evaluation not relying on self-report could supply an important addition especially in non-communicable patients and, relatedly, a neural signature for pain in adults has been proposed (Wager et al., 2013) and debated (Jaillard and Ropper, 2013; Lu et al., 2013; Apkarian, 2013). Its relevance in clinical practice remains to be evaluated. Importantly, this method seeks to bypass self-report. The ambition here, on the contrary, is to analyse the possibility of augmenting the capacity to self-report by means of neuro-technological communication (NTC).

Owen and colleagues demonstrated awareness in an otherwise non-responsive patient suffering from traumatic brain injury (Owen et al., 2006). On clinical assessment, the criteria for vegetative state (preserved arousal without awareness) were fulfilled. By means of neuro-imaging, intended purposeful behaviour was nevertheless demonstrated in relation to spatial imagery tasks; discrete neural activations indistinguishable from those in healthy controls were obtained during instructions to imagine playing tennis and perform a spatial navigation task indicating conscious awareness of self and surroundings. Moreover, it was suggested that reproducible robust task-dependent responses to command would supply a method by which otherwise unresponsive patients with preserved awareness could communicate. This paradigm case (henceforth so referred) introduced the concept of NTC illustrating the feasibility of augmenting the capacity to self-report in otherwise nonresponsive patients. Prime candidates include patients suffering from Disorders of Consciousness (DoC) whose autonomy may benefit significantly from a means to communicate.

Clinical assessment of pain in preverbal patients, typically neonates, is performed by observation of behaviour (facial expression, crying, positioning, movement, etc.) and physiological parameters (changes in heart rate, respiratory rate, blood pressure, and oxygen saturation) (Maxwell et al., 2013). The parameters being unspecific and particularly sensitive to distress in general, assessment is difficult and uncertain. Also, neither in paediatric nor in neonatal practise are disorders affecting brain function and consciousness rarities (Duhaime and Rindler, 2015). These patients would be likely to benefit from an expanded arsenal of communication.

Attributing consciousness

Preverbal infants signal their needs effectively; an array of evolutionary conserved behaviours (Panksepp, 2005) secures support and rearing crucial to survival for individual and species alike. The range of communicable content is limited nevertheless adapted to actual needs. Until recently, infant behaviour was to a large extent considered the outcome of reflexes and medical procedures were conducted accordingly, anaesthesia and pain relief only infrequently administered (Anand, 2007). At present day, however, we generally assert consciousness on the basis of such behaviour (Lagercrantz and Changeux, 2010; Lagercrantz, 2014).

A behavioural response, like that in the paradigm case, implicates intact sensory pathways and information processing at some level *corresponding* to that of a conscious subject. However, there exists, strictly, no direct evidence for the response behaviour's correspondence to consciousness[1]. The inference from neuro-technologically recognised behaviour to consciousness *being the best explanation* is not equivalent to an inference from neuro-technologically recognised behaviour to consciousness *being the explanation*. Notably this is not a problem unique to NTC.

In any communicative interaction an evaluation of the counterpart's properties and an associated judgement as to whether or not to ascribe to the counterpart a conscious mind is performed, although typically intuitively. In lack of an empirical method other than evaluation of behaviour, the conditions under which we ascribe consciousness to someone or something are either, in everyday practice, intuitive, or, in a more stringent context, postulated, and importantly, fallible. Consequently, and in particular when overt behavioural signals are difficult to interpret or absent, as may be the case in preverbal infants and unresponsive patients respectively, an ascription of consciousness is just that[2].

At the heart of the scepticism towards ascribing consciousness based on behaviour lies the lack of a theory providing an explanatory account of the mechanism linking third-person observed brain-readings to first-person accounts of *what it is like* (Nagel, 1974) to be conscious. Available accounts fail to bridge the gap between phenomenal experience and the models provided by empirical science. The contemporary definition of the problem allegedly emanates in Descartes' *Second Meditation* in which mind and body are separated (Descartes, 2008). Analyses of the problem are known by catchphrases; "The hard problem of consciousness" (Chalmers, 1996) (how to explain phenomenal consciousness) or "The explanatory gap" (Levine, 1983) (pain spelled out as "c-fibers firing" does not help us understand what pain feels like). A theory bridging the gap, however distant it may seem (Nagel, 2012), would provide the basis for physico-psychological inferences. Meanwhile, attributing consciousness will be liable to critique from this angle.

Despite this fundamental methodological obstacle, the correspondence between external behaviour and intuitively inferred mental states is typically taken to imply conscious awareness (Lagercrantz and Changeux, 2010) and indeed a self, also in preverbal infants (Rochat, 2011).

Possible targets

In the paradigm case cortical activity evoked through language cues is exploited. NTC so understood relies on sensory input, higher cognitive processing involving language comprehension, volition, and executive function. In preverbal infants, language absence renders verbal behavioural cues, like those exploited in the paradigm case, blunt.

The brain is operational during development and the infant's capacity to interpret signals as well as to transmit them is consequently rapidly changing (Rochat, 2003, 2011). Thus, brain development restricts and enables the neurophysiological correlates available in NTC efforts. Thalamocortical connections proposed necessary and sufficient for basic consciousness are present from 24 weeks of gestation (Lagercrantz and Changeux, 2010) and at two months of age perceptual categorisation enabling concept formation has been demonstrated (Westermann and Mareschal, 2014). At birth infants discriminate between self and world; at two months a sense of own agency is manifest; at four months the capacity of hand-eye coordination is present; at 18 months self-consciousness is manifest (Rochat, 2003, 2011). Underlying these developmental stages are the maturation of different brain systems. The successive operation of orbitofrontal, venterolateral, dorsolateral, and rostrolateral cortices have been suggested to imply particular levels of cognition including executive function and rule-use (Rochat, 2011). However, the developmental stage at which cognitive processing supports the ability to receive and transmit signals is currently obscure. And, assuming that ontogenetic development follows phylogeny, other systems than those involved in cognition may supply appropriate targets for NTC in preverbal infants.

Relatedly, next to the substantial literature in the field of cognitive neuroscience[3], affective mental states and correlated neural mechanisms attract growing attention (Panksepp, 2005, 2003; Denton et al., 2009; Merker, 2007). Proponents conceive of the young nervous system as subserving affective aspects of consciousness responsible for behaviour whereas cognition remains to develop with the expansion and maturation of the rostral end of the neuro-axis.

Panksepp holds affective states intrinsically motivating and dictated by evolution – "an unconditional gift of nature" (Panksepp, 2005) – providing not only the basic "energizer" that drives ecological behaviour, but also the feeling aspect of experience (qualia). Affective states are considered prior to cognition, from the phylogenetic as well as the ontogenetic perspective, and shared by all mammals in virtue of subcortical localisation.

Denton introduces the concept of *primordial emotion* amounting to "an imperious specific sensation and the compelling intention" yet lacking a cognitive component, and postulates it to be the evolutionary beginning of consciousness and, ontogenetically, of the organism's conscious awareness (Denton et al., 2009).

Merker proposes an ontogenetic and phylogenetic early *analog reality simulation* incorporating target selection, action selection, and motivation; to be operating in the dorsal midbrain by which the organism's fundamental requirement of matching opportunities, behaviour and basic needs is met. While the midbrain system supports immediate unreflective experience – *primary consciousness* – at the core of consciousness, more sophisticated aspects such as self-consciousness are argued to be products of the neocortex (Merker, 2007).

The preverbal brain is, following the accounts of Panksepp, Denton, and Merkel, unlikely to answer the call on a cognitive level. However, by targeting affective states, preverbal infants could be invited to the conversation. Consequently, it will be argued that examining the possibility of augmenting the capacity to self-report in preverbal infants through NTC will prompt an analysis of affective states. However, first, in order to evaluate the success of NTC efforts in preverbal infants, its real world counterpart – ordinary infant communication – will be examined.

Communication with preverbal infants

Communication, it may be suggested, is the successful transfer of a message from sender to recipient if the message sent is also that received, and if the parties are convinced that so is the case. The latter is typically the result of corrective inputs generating *convergence* on the subject matter; the looping process of communication is self-stabilising and converges on a joint understanding of the content communicated. Arguably, a minimum criterion for a successful communicative process is the possibility to generate convergence by corrective inputs.

In the paradigm case, communication is made possible by representative coding. The attribution of consented arbitrary meaning to neurophysiological activity or behaviour amounts to the construction of a representative language, which allows consistent reproduction of responses signifying something other than the neural activation itself. Thus, the activity in premotor areas doesn't denote "playing tennis"; it functions in response to actual questions much like the eye-blink of a patient with locked-in syndrome. Apart from enabling convergence by supporting corrective signals and reproducibility, attribution of representations to proxies permits versatility and thus in principle endless diversity of communicated content. Additionally, and importantly, the resulting behaviour warrants an attribution of consciousness.

Examining pain processing in neonates, haemodynamic changes associated with activation of primary somatosensory cortex were measured and demonstrated to correlate to painful stimulus (heel prick venepuncture) (Bartocci et al., 2006). The authors take the activity to imply conscious sensory perception. Could such evoked activity function in NTC? On what analysis would it amount to a self-report?

In comparison to the paradigm case, a different capacity to meet the criterion of convergence is evident. An evoked physiological parameter beyond volitional control, e.g., somatosensory activity in response to a noxious stimulus, can only denote what it is. It cannot serve as to represent anything else and it will never confirm or refute an interpretation of the signal. Correction signals, and convergence thereby, are impossible as all there is, is neural activity in response to a noxious stimulus.

Further, the incapacity to select attributions substantially limits what can be communicated; no versatility is possible. In the adult case this would amount to a tedious conversation involving "playing tennis" and "spatial navigation", and only that. The difference between measuring evoked physiological parameters on the one hand, and evoked physiological parameters being proxies for "yes" or "no" on the other, is substantial and impacts on versatility, range, and convergence.

However, different standards are conceivably appropriate in the infant case as normal communicative interaction implies. With respect to versatility, the infant's repertoire of mental states is limited (Rochat, 2011; Lagercrantz, 2014) and representational coding adding value consequently improbable. Further, in ordinary communication with infants it is clear that convergence is possible, however, over the limited range of states and by a different set of corrective signals consisting largely in negative and positive feedback cues. In short, infants don't rephrase, they persist, and we keep guessing until they are content.

The content of infant self-reports is presumably restricted. Consequently, even though an evoked potential reflecting an internal state in preverbal infants only denote that state, the possibility of evaluating it in relation to manipulation (feeding, warming, comforting, etc.) could prove meaningful. And, indeed, in relation to its real world counterpart, it could be interpreted as a self-report.

However, corrective signals and thus convergence being impossible, successful communication would be crucially dependent on the interpretation being correct (i.e., the evoked potential actually corresponding to the interpreted state) as an evoked response can be neither confirmed nor refuted – only observed to vary – in relation to the interpretation. Consequently, uncertainty and fallibility are anticipated. Nevertheless, this is in principle also the case in ordinary communication with preverbal infants.

The success of NTC in augmenting self-reports of preverbal infants need not then, considering its real world counterpart, be evaluated in relation to criteria applicable in the paradigm case. Evoked brain activity corresponding to conscious states could be interpreted analogously to ordinary infant behaviour. The capacity to function as self-reports in successful communication relies on the message sent also being that received and the degree of certainty the parties attribute to it being so. If, from the nociceptive processing it displays, we conceive of an infant as being in pain, this amounts to a self-report and, "communication" can be accurately ascribed to that process.

Conceivably, detectable neural activity signalling specific affective states corresponding to the repertoire of uniform infant behaviour supply communicable content. Also this repertoire together with its corresponding neural systems being evolutionary conserved, it will be argued an opportunity to attain knowledge from the study of other mammals is available. An already suggested trans-species approach (Panksepp, 2011) can direct the interpreter in decoding brain activity and eventually in formulating a theory guiding physico-psychological inferences.

Neural infantese

The newborn infant's behavioural repertoire functions among other things to signal distress and elicit rearing. Infant generated cues exhibiting cross-species effects (Parsons et al., 2011; Yong and Ruffman, 2014) suggests evolutionary conserved mechanisms underlying these behaviours which invites the introduction of "infantese" as a term denoting that array of behaviours. Further, infant response to baby-speech – "motherese" – initially indiscriminate to human and non-human primate vocalisations (Ferry et al., 2013), nevertheless, with effects on cognitive and social development (Saint-Georges et al., 2013), imply an ecologically important conserved mechanisms underlying also infant response behaviour. Evolutionary conserved behavioural patterns thus appear to govern early life self-reporting and responding. Augmentation of self-reports in preverbal infants by NTC, could rely on neural derivatives of infantese provided these are parts of conscious processing. What neural activity corresponds to conscious processing in the preverbal infant, and can it subserve infantese?

Most contemporary analyses of consciousness focus on its cognitive aspects and the cerebral cortex. An extensive body of literature exists linking cortical activity to conscious processing. Evidence emanates largely from verbal or other behavioural reports of conscious experience. In essence the experiments contrast physiological mechanisms involved in conscious sensory perception to those in minimally different yet non-conscious processing. *Reportability*, not only a method but also a criterion for consciousness research on this construal (Baars, 2005; Dehaene and Changeux, 2011; Tononi and Koch, 2008), excludes non-verbal infants as well as most animals although the consensual agreement, or the general intuition, respectively, is that these are indeed conscious. Also, comparison of neural activity in the awake conscious state from that in other states where consciousness is void such as sleep (Steriade, 2003), anaesthesia (Barttfeld et al., 2015), epileptic states of absence (Englot and Blumenfeld, 2009), and disorders of consciousness (Vanhaudenhuyse et al., 2010) have been exploited.

On the basis of data obtained by these approaches (with rare exceptions [Kouider et al., 2013] unexplored in preverbal infants), dynamic properties of the summation of cortical discharges – sustained or phasic activity, re-entrant or feed-forward activity, synchronised and oscillatory activity – have been proposed to correspond to the neural correlates of consciousness (Tononi and Koch, 2008) and attempts at creating theoretical models to describe, or even explain, consciousness on a cortical level have been made including the global workspace theory (Baars, 2005) and the integrated information theory (Tononi and Koch, 2008).

Such accounts have met criticism claiming them to describe information processing rather than phenomenal consciousness. In particular, affective states, marked by phenomenal qualities, central not only to the analysis of motivational impetus but to that of phenomenal consciousness in general, arguably resist analysis on an altogether cognitive cortical model of consciousness (Block, 2009). Applying to augment the capacity to self-report by exploiting neural substrates correlated to infantese, which it was argued, corresponds to internal affective, i.e., conscious states, NTC with preverbal infants hinges on the substrate being conscious, and thus it is inconceivable in relation to brain activity unlikely to correspond to such states.

Also, several lines of evidence imply cortical activity not being necessary for conscious: anencephalic children exhibit behaviour interpreted as to indicate consciousness awareness (Merker, 2007); removal of large cortical sectors, including hemispherectomy, in human awake subjects, failed to impair consciousness per se (Penfield, 1954); decorticated animals (rats, cats) exhibit complex behaviour indicative of conscious awareness (Merker, 2007); absence epilepsy, marked by lapse of consciousness, failed to be triggered by cortical stimulation as opposed to all other kinds of seizures (Penfield, 1954),[4] and; neither damage nor stimulation to somatosensory cortex inflict pain (Craig, 2003a).

Cortical activity, linked to conscious experience by methods relying on reportability, is at the very least insufficient, and possibly even unnecessary, for consciousness as such, and in preverbal infants, where most of the cognitive capacity is yet to develop, a cortical approach is even more unlikely to generate in success.

Consciousness arguably ecologically relevant in virtue of promoting self-perseverance through matching internal biological needs to external opportunities, and eliciting motives for, and ultimately actions aimed at, fulfilling these needs; infants' first conscious states are likely to reflect this conglomerate of biological need/motive/opportunity/action. Infantese presumably answering to core homeostasis as well as being conserved in mammalian species, the neural correlates of these conglomerates are likely to mature early and be evolutionarily conserved. Consequently, its anatomical basis is likely to involve ancient parts of the nervous system.

Analyses taking this approach have been suggested (Denton et al., 2009; Merker, 2007; Panksepp, 2005; Northoff and Panksepp, 2008; Craig, 2003b), however, not pertaining to infant consciousness in particular. Nevertheless, unifying among these are; (1) the integration of interoceptive stimuli from (a) bodily states, (b) intrinsic brain states of attentional, emotional, and/or motivational character, and (c) environmental exteroceptive stimuli, into a goal-orientated self-preserving system; (2) subcortical structures are identified as key components; (3) these are preserved across mammalian species; and (4) supply a primal form of self-processing/representation, a self.

On Denton's construal (Denton et al., 2009), *primordial emotions*, "the subjective elements of instincts", comprising of thirst, hunger, pain, etc., subserve the vegetative systems to "guard the physico-chemical constancy of the internal environment of the body". A specific subjective sensation paired with compelling intention evolved early in phylogeny as a powerful mechanism signalling not only a threat to the organism but also a reason to act, thereby supporting survival. "The earliest dim awareness – the initial subjectivity of consciousness", an atom of consciousness, is proposed to comprise of the conglomerate of sensation and intention which is referred to as a "primordial emotion". Indeed, recent findings from neuroimaging studies in adults demonstrate mechanisms subserving primordial emotions, including thirst, hunger, hunger for air, pain, and visceral distension to involve structures such as anterior, middle, and posterior cingulate areas and parahippocampal gyri, insula, claustrum, thalamus, amygdala, diencephalon, and midbrain (thirst); and, predominantly insula and limbic system (pain) (Denton et al., 2009).

Merker (2007), updates the *centrencephalic system* proposed by Penfield and Jaspers, who through the removal of cortical sectors, including hemispherectomy, in awake patients suffering from severe epilepsy, noted, from self-report during surgery, no deprivation of consciousness per se, but rather of certain forms of information or capacities, leading to the proposal that subcortical mechanisms are *functionally supra-cortical* in terms of conscious volition. Merker recognises the organism's need for organising triangular dependency between motivation, target selection, and action selection in order to control behaviour hence solving decision problems, ultimately in relation self-perseverance subject to evolutionary principles. An integrative machinery within the mesodiencephalon (midbrain and diencephalon) consisting of (1) the hypothalamus, subserving goal-directed motivated behaviour, (2) the superior colliculus, where spatial senses are superposed within a common premotor framework for orienting, and (3) projections from the substantia nigra and intermediate collicular layers, contributing basal ganglia motor action information, form a *selection triangle* presenting a solution to the decision problem through an *analog reality simulation*. Merker proposes this arrangement to have "served as the innate scaffolding supporting all further elaboration of conscious contents in phylogeny" (Merker, 2007).

Northoff and Panksepp (2008) propose a "trans-species concept of self", involving a "sensorimotor, valuative, intero-exteroceptive and subjective-affective experience integrator" enabling self-related processing, residing in the subcortical-cortical midline system homologous across mammalian species. They recognise a comprehensive network of brain regions including, among other components, the periaqueductal grey (PAG) and adjacent tectal maps being responsible for convergence of emotional processing and relation to bodily and environmental stimuli while reward seeking – the evaluation of stimuli in relation to organism's needs – is supported in ventral tegmental area (VTA) with projections into ventral striatum (VS) and nucleus accumbens (NACC) together with various other sub-cortical structures such as hypothalamus, amygdala, and dorsomedial thalamus as well as medial prefrontal cortices (mPPC). Northoff and Panksepp also recognise experience to epigenetically drive neocortical maturation as to account for a number of cognitive functions while self-related processing is suggested to be sustained by a subcortical homologue.

Empirical findings supporting the relevance of subcortical structures for consciousness, other than surgical removal of cortex (Penfield, 1954), consist, among others, of: neocortical brain stimulation resulting in substantially less intense affective behaviour than stimulation in subcortical regions (Panksepp, 2005); damage to subcortical substrates (periaqueductal grey matter and medial and intralaminar thalamic nuclei) impair consciousness dramatically (as opposed to in neocortical regions) (Watt and Pincus, 2004), and; decorticated animals and anencephalic children manifest behaviour that indicates phenomenal consciousness (Merker, 2007). Also, in a recent PET-study of subjects returning from anaesthesia, activity in the locus coeruleus, hypothalamus, thalamus, and anterior cingulate was primarily correlated with the emergence of consciousness while neocortical involvement sparse (Långsjö et al., 2012).

The content of consciousness could be conceived of as developing from primordial emotions or an analog reality simulation subserved in evolutionary conserved parts of the brain, i.e., the brain stem and midline structures, to later incorporate cognitive capacities expanding the phenomenological repertoire and, concomitantly, survival fitness. Such non-cognitive basic states of self-awareness, motivation, and intention are conceivably subserved in subcortical substrates.

In preverbal infants, with a cognitive capacity only starting to develop, and a nervous system conceivably resembling that of earlier phylogeny; however, plausibly exhibiting operational subcortical structures, and an evolutionary conserved range of behaviours, deriving from internal conscious states involving sensory affective intentions and motor control subserved on a brainstem level, and related to basic homeostatic needs, it is suggested, that the corresponding neural activity could by augmentation through NTC generate in detectable self-reports; or, "neural infantese".

Detecting pain

Premature infants exhibit coordinated facial movements indicative of pain in response to heel lance (Craig et al., 1993) and from 25 weeks of gestation preterm infants display somatosensory cortex activation measured by near infrared spectroscopy (NIRS) under the same condition (Bartocci et al., 2006) interpreted as to indicate pain. Further, fMRI at full term reveal distinct activity patterns in relation to modalities of nociceptive and tactile stimuli as well as to intensity (Williams et al., 2015) suggesting a cortical discrimination of these modalities.

These experiments demonstrate cortical processing in response to noxious stimulus; however, the relation to conscious experience of pain is speculative (which is not to say pain is not present; however, cortical activity is not a reliable indication that so is the case). As noted previously, neither damage to, nor stimulation of somatosensory cortex, inflict pain (Craig, 2003a). Further, nociceptive processing, traditionally conceived of as supported by peripheral nociceptors, lamina I of the dorsal horn, the contralateral spinothalamic tract and thalamocortical projections to the somatosensory cortex triggering a conscious perception of pain, is outdated. The current general understanding (in adults) identifies a multifocal "pain matrix" comprising of lateral (sensory-discriminatory) and medial (affective-cognitive-evaluative) neuroanatomical components under constant and active modulation of a variety of brain regions including in particular the brain stem (Tracey and Mantyh, 2007). Encompassing sensory physiological response to noxious stimulus as well as affective experience and cognitive conceptualisation – the later two both influenced by memory, mood, attention, and context – nociceptive processing involve primary and secondary somatosensory cortices, insular cortex, anterior cingulate cortex (ACC), prefrontal cortices and the thalamus as well as basal ganglia, cerebellum, amygdala, hippocampus, and areas within the temporal and parietal lobes (Tracey and Mantyh, 2007). In particular the ACC has been implicated in processing of the affective component of pain (Fuchs et al., 2014) and, the right (non-dominant) insular cortex in the representation of pain (Craig, 2003b). An attribution of pain merely from activity in somatosensory cortex is problematic, however, early in life, not only in relation to the recent reconceptualisation of pain processing invoking the pain matrix.

Already before term, the anatomical substrates for transmission of somatosensory stimuli to the developing cortex are present (Lagercrantz and Changeux, 2010). It may be proposed that once these are in place, other critical stages, demarking the transition from sensing to feeling (Rochat, 2011) and concomitant emergence of self-awareness, are difficult to define. Consequently, in relation to pain, it may be hypothesised that once somatosensory areas exhibit activity on noxious stimulation, activity is present in the pain-matrix as a whole.

However, prior to 26 weeks of gestation thalamocortical pathways reach at most into the subplate of the developing cortex and activation of these pathways result primarily in a widespread prolonged electrical excitation (slow activity; 100–1000 ms) of the subplate layer which subsequently may induce activity in the overlaying cortical layer (Vanhatalo et al., 2009). From 26 to 32 weeks the thalamocortical fibers reach deeper cortical layers and a gradual change from slow to fast oscillations (down to 1/10 ms) commence. At full term the slow activity has disappeared indicating its origin in transient structures and so less likely to support self-awareness.

The slow activity has been suggested to emanate from endogenously generated intermittent "pacing" of subcortical elements in the developing nervous system essential for survival of neurons (Hebb's postulate) and the development of long-range connectivity (Vanhatalo and Kaila, 2006), arguably important to cognitive processing (Dehaene and Changeux, 2011; Barttfeld et al., 2015), and as such they are less likely to support meaningful experience.

The ensuing fast activity is a result of intracortical transmissions enabled chiefly by interneuron growth and maturation; it is the normal activity of full term babies and older subjects and meet the criteria of integration and differentiation argued to support cognitive processing (Tononi and Koch, 2008). However, the functional architecture in infants supports tasks of action-perception nature thereby differing from the adult where associative cortices dominate; the adult somatosensory homunculus need not then translate to that of the infant (Fransson et al., 2011).

The maturing brain is operational while in development. Nevertheless, in early life, the dynamics of cortical activity, as well as the neurobiological function it is suggested to play in relation to development, indicate subjective experience arising from cortical activity unlikely, and even if so, difficult to interpret. If infants are indeed considered capable of perceiving pain, evoked cortical responses are not reliable indicators thereof. This, importantly, does not contradict painful experience, which well may be subserved in other brain regions. Instead, and in analogy with what had been argued in the previous, it emphasises the need to explore the infant brain for activity that can function as self-reports.

Craig notes that in non-primates, lamina I activity is propagated and integrated on a brainstem level while, in primates, there is also a direct thalamocortical homeostatic afferent pathway projecting to the dorsal posterior insular cortex to provide discrete topographic modality specific representations of interoceptive lamina I input such as sharp pain, burning pain, cool, warm, itch, sensual touch, and muscle burn. Together with parasympathetic afferents from the solitary nucleus, supplying input vis-à-vis hunger, thirst, cardiorespiratory, etc., this pathway is suggested to provide an "encephalised" metarepresentation of the physiological state of the body proposed to be associated with "subjective awareness of the material self as a feeling (sentient) entity" (Craig, 2003a). In newborn infants insular and ACC activity was demonstrated in response to tactile stimulus (Williams et al., 2015) providing preliminary support for a feasible extrapolation of Craig's analysis to preverbal infants.

Damasio conducted explorations into the subcortical and cortical brain activity during the feeling of self-generated emotions. The demonstration of discrete neural patterns constituting multidimensional maps of the organism's internal state in relation to sadness, happiness, anger, and fear was proposed to supply the functional substrate for feelings. Anger, for instance, activated midbrain and pons along with anterior half of left cingulate while bilaterally deactivating secondary somatosensory cortex (Damasio et al., 2000). Pertaining to adult subjects, these data encourage elaboration in preverbal infants.

Denton reviews studies of neural activity in relation to the proposed primordial emotions of thirst, hunger, "hunger for air", pain and visceral distension and finds evidence that in adult subjects, structures of both brain stem and telencephalon play a dominant role (Denton et al., 2009).

Future directions

Pain in preverbal infants is behaviourally indicated (Craig et al., 1993) and consensually agreed to be manifest. The relevance of approaches relying on cortical activity remains to be evaluated. However, here it has been proposed that in preverbal infants, conceivably exhibiting operational subcortical structures and an evolutionary conserved range of behaviours, deriving from internal conscious states involving sensory affective intentions and motor control, subserved on a brain stem level, and related to basic homeostatic needs, the corresponding neural activity could, by augmentation through NTC generate in detectable self-reports corresponding to infantese.

Knowledge of human brain development is rapidly expanding and will undoubtedly contribute to furthering our understanding of the developing brain. The proposal here put forward relies on understanding evolutionary conserved brain systems and reaction patterns displaying cross-species similarities (Panksepp, 2005) which emphasises the need for trans-species integrative research (Panksepp, 2011). By relating human self reports of basic emotions, such as pain, and evolutionary conserved reaction patterns, to those of animals exhibiting behaviour that indicate a similar internal state or behaviour, underlying neural systems can be decoded. This triangulation approach has been proposed (Panksepp, 2005; Panksepp and Northoff, 2009; Northoff and Panksepp, 2008) however not in infants and not in the present context. It may guide in formulating a theory of physico-psychological inference.

Conclusion

Efforts at NTC with preverbal infants need differ from the paradigm exploited in adult subjects. Adhering to limitations in language capacity, arbitrary meaning attribution is inconceivable. Brain states directly representing conscious awareness could instead supply substrates for communication. In relation to empirical evidence of neural substrates of consciousness, and with particular regards to early developmental level of preverbal infants, these states are proposed to emanate primarily from evolutionary conserved structures. By exploring neural substrates underlying behaviour present across species and relating it to human emotional states and behaviour, such as pain and pain-related behaviour, a triangulation approach may prove valuable in decoding the neurobiology of early life pain and suffering. Regardless of the results from such endeavours, conscious experience, e.g., pain, is only fully appreciated from the first-person perspective; importantly, this needs to be kept in mind regardless of developmental level.

Acknowledgments

I thank Hugo Lagercrantz, Olof Lagerlöf, and Predrag Petrovic for valuable comments on an early draft of this chapter.

Notes

1 This point is made by Chalmers in a series of thought experiments involving zombies, arguably lacking conscious experience yet behaviourally identical to normal persons (Chalmers, 1996).

2 In fact, in line with what has been argued, this would be generally true. See, e.g., Miedaner (Miedaner, 1981).
3 Cognitive neuroscience typically engage perception, attention, memory, knowledge representation and categorisation, language and thinking (including reasoning and decision making) (Galotti, 2011). Affective neuroscience deals with affective aspects of consciousness suggested to impact on cognitive, behavioural, expressive and physiological processes (Panksepp, 2005). The division in cognitive and affective neuroscience is not undisputed (Panksepp, 2003; Storbeck & Clore, 2007).
4 Instead it was triggered by stimulating the midline thalamus in cats and later found linked to the zona incerta (Merker, 2007).

References

Anand, K. J. S. 2007. Consciousness, cortical function, and pain perception in non-verbal humans. *Behavioral and Brain Sciences*, 30(1), 82–3.

Apkarian, A. V. 2013. A brain signature for acute pain. *Trends in Cognitive Sciences*, 17(7), 309–10.

Baars, B. J. 2005. Global workspace theory of consciousness: Toward a cognitive neuroscience of human experience. *Progress in Brain Research*, 150, 45–53.

Bartocci, M., Bergqvist, L. L., Lagercrantz, H., & Anand, K. J. 2006. Pain activates cortical areas in the preterm newborn brain. *Pain*, 122, 109–17.

Barttfeld, P., Uhrig, L., Sitt, J. D., Sigman, M., Jarraya, B., & Dehaene, S. 2015. Signature of consciousness in the dynamics of resting-state brain activity. *Proceedings of the National Academy of Sciences*, 112(3), 887–92.

Block, N. 2009. Comparing the major theories of consciousness. In Gazzaniga, M. (Ed.), *The Cognitive Neurosciences*, 1111–22.

Chalmers, D. J. 1996. *The conscious mind: In search of a fundamental theory.* New York and Oxford: Oxford University Press.

Craig, A. D. 2003a. A new view of pain as a homeostatic emotion. *Trends in Neurosciences*, 26(6), 303–7.

Craig, A. D. 2003b. Interoception: The sense of the physiological condition of the body. *Current Opinion in Neurobiology*, 13(4), 500–5.

Craig, K. D., Whitfield, M. F., Grunau, R. V., Linton, J., & Hadjistavropoulos, H. D. 1993. Pain in the preterm neonate – behavioral and physiological indexes. *Pain*, 52, 287–99.

Damasio, A. R., Grabowski, T. J., Bechara, A., Damasio, H., Ponto, L. L., Parvizi, J., & Hichwa, R. D. 2000. Subcortical and cortical brain activity during the feeling of self-generated emotions. *Nature Neuroscience*, 3(10), 1049–56.

Dehaene, S., & Changeux, J. P. 2011. Experimental and theoretical approaches to conscious processing. *Neuron*, 70(2), 200–27.

Denton, D. A., Mckinley, M. J., Farrell, M., & Egan, G. F. 2009. The role of primordial emotions in the evolutionary origin of consciousness. *Consciousness and Cognition*, 18(2), 500–14.

Descartes, R. 2008. *Descartes: Meditations on first philosophy.*

Duhaime, A. -C., & Rindler, R. S. 2015. Special considerations in infants and children. *Handbook of Clinical Neurology*, 127, 219–42.

Englot, D. J., & Blumenfeld, H. 2009. Consciousness and epilepsy: Why are complex-partial seizures complex? *Progress in Brain Research*, 177, 147–70.

Ferry, A. L., Hespos, S. J., & Waxman, S. R. 2013. Nonhuman primate vocalizations support categorization in very young human infants. *Proceedings of the National Academy of Sciences of the United States of America*, 110(38), 15231–5.

Fransson, P., Aden, U., Blennow, M., & Lagercrantz, H. 2011. The functional architecture of the infant brain as revealed by resting-state fMRI. *Cerebral Cortex*, 21, 145–54.

Fuchs, P. N., Peng, Y. B., Boyette-Davis, J. A., & Uhelski, M. L. 2014. The anterior cingulate cortex and pain processing. *Frontiers in Integrative Neuroscience*, 8, 35.

Galotti, K. 2011. Cognitive development: Infancy through adolescence. *Cognitive Development: Infancy through Adolescence*, 335–72.

Jaillard, A., & Ropper, A. H., 2013. Pain, heat, and emotion with functional MRI. *The New England Journal of Medicine*, 368(15), 1447–9.

Kouider, S., Stanhult, C., Gelskov, S. V., Barbosa, L. S., Dutat, M., De Gardelle, V., Christophe, A., Dehaene, S., & Dehaene-Lambertz, G. 2013. A neural marker of perceptual consciousness in infants. *Science*, 340(6130), 376–80.

Lagercrantz, H. 2014. The emergence of consciousness: Science and ethics. *Seminars in Fetal & Neonatal Medicine*, 19(5), 300–5.

Lagercrantz, H., & Changeux, J. P. 2010. Basic consciousness of the newborn. *Seminars in Perinatology*, 34, 201–6.

Långsjö, J. W., Alkire, M. T., Kaskinoro, K., Hayama, H., Maksimow, A., Kaisti, K. K., Aalto, S., Aantaa, R., Jääskeläinen, S. K., Revonsuo, A., & Scheinin, H. 2012. Returning from oblivion: Imaging the neural core of consciousness. *The Journal of Neuroscience: The Official Journal of the Society for Neuroscience*, 32(14), 4935–43.

Levine, J. 1983. Materialism and qualia: The explanatory gap. *Pacific Philosophical Quarterly*, 64, 354–61.

Lu, Y., Klein, G. T., & Wang, M. Y. 2013. Can pain be measured objectively? *Neurosurgery*, 73(2), 24–5.

Maxwell, L. G., Malavolta, C. P., & Fraga, M. V. 2013. Assessment of pain in the neonate. *Clinics in Perinatology*, 40(3), 457–69.

Merker, B. 2007. Consciousness without a cerebral cortex: A challenge for neuroscience and medicine. *The Behavioral and Brain Sciences*, 30, 63–81; discussion 81–134.

Merskey, H., & Bogduk, N. (Eds.). 1994. *Classification of chronic pain. Descriptions of chronic pain syndromes and definitions of pain terms*. Seattle: IASP Press.

Miedaner, T. 1981. The soul of the Mark III beast. In D. R. Hofstadter & D. C. Dennett, eds., *The Mind's I. Fantasies and Reflexions on Self and Soul*, pp. 109–13. New York: Bantam Books.

Nagel, T. 1974. What is it like to be a bat. *Philosophical Review*, 83, 435–50.

Nagel, T. 2012. *Mind and cosmos : Why the materialist neo-Darwinian conception of nature is almost certainly false*. New York: Oxford University Press.

Northoff, G., & Panksepp, J. 2008. The trans-species concept of self and the subcortical-cortical midline system. *Trends in Cognitive Sciences*, 12(7), 259–64.

Owen, A. M., Coleman, M. R., Boly, M. H., Davis, M. H., Laureys, S., & Pickard, J. D. 2006. Detecting awareness in the vegetative state. *Science*, 313(5792), 1402.

Panksepp, J. 2003. At the interface of the affective, behavioral, and cognitive neurosciences: Decoding the emotional feelings of the brain. *Brain and Cognition*, 52(1), 4–14.

Panksepp, J. 2005. Affective consciousness: Core emotional feelings in animals and humans. *Consciousness and Cognition*, 14(1), 30–80.

Panksepp, J. 2011. Cross-species affective neuroscience decoding of the primal affective experiences of humans and related animals. *PloS ONE*, 6(9), e21236.

Panksepp, J., & Northoff, G. 2009. The trans-species core SELF: The emergence of active cultural and neuro-ecological agents through self-related processing within subcortical-cortical midline networks. *Consciousness and Cognition*, 18(1), 193–215.

Parsons, C. E., Young, K. S., Pearsons, E., Dean, A., Murray, L., Goodacre, T., Dalton, L., Stein, A., & Kringelbach, M. L. 2011. The effect of cleft lip on adults' responses to faces: Cross-species findings. *PLoS ONE*, 6(10), 1–6.

Penfield, W. 1954. *Epilepsy and the functional anatomy of the human brain*, London: Churchill.

Price, D. D. 2000. Psychological and neural mechanisms of the affective dimension of pain. *Science*, 288(5472), 1769–72.

Rochat, P. 2003. Five levels of self-awareness as they unfold early in life. *Consciousness and Cognition*, 12(4), 717–31.

Rochat, P. 2011. The self as phenotype. *Consciousness and Cognition*, 20(1), 109–19.

Saint-Georges, C., Chetouani, M., Cassel, R., Apicella, F., Mahdhaoui, A., Muratori, F., Laznik, M. C., & Cohen, D. 2013. Motherese in interaction: At the cross-road of emotion and cognition? (A systematic review). *PloS ONE*, 8(10), e78103.

Steriade, M. 2003. The corticothalamic system in sleep. *Frontiers in Bioscience: A Journal and Virtual Library*, 8, d878–99.

Storbeck, J., & Clore, G. L. 2007. On the interdependence of cognition and emotion. *Cognition & Emotion*, 21(6), 1212–37.

Tononi, G., & Koch, C. 2008. The neural correlates of consciousness: An update. *Annals of the New York Academy of Sciences*, 1124, 239–61.

Tracey, I., & Mantyh, P. W. 2007. The cerebral signature for pain perception and its modulation. *Neuron*, 55(3), 377–91.

Vanhatalo, S., Jousmäki, V., Andersson, S., & Metsäranta, M. 2009. An easy and practical method for routine, bedside testing of somatosensory systems in extremely low birth weight infants. *Pediatric Research*, 66, 710–13.

Vanhatalo, S., & Kaila, K. 2006. Development of neonatal EEG activity: From phenomenology to physiology. *Seminars in Fetal & Neonatal Medicine*, 11(6), 471–8.

Vanhaudenhuyse, A., Noirhomme, Q., Tshibanca, L. J., Bruno, M. A., Boveroux, P., Schnakers, C., Soddu, A., Perlbarg, V., Ledoux, D., Brichant, J. F., Moonen, G., Maquet, P., Greicius, M. D., Laureys, S., Boly, M. 2010. Default network connectivity reflects the level of consciousness in non-communicative brain-damaged patients. *Brain*, 133(2009), 161–71.

Wager, T. D., Atlas, L. Y., Lindquist, M. A., Roy, M., Woo, C. -W., & Kross, E. 2013. An fMRI-based neurologic signature of physical pain. *The New England journal of medicine*, 368(15), 1388–97.

Watt, D. F., & Pincus, D. I. 2004. Neural substrates of consciousness: Implications for clinical psychiatry. In J. Panksepp (Ed.), *Textbook of biological psychiatry*, pp. 75–110. Hoboken, NJ: John Wiley & Sons.

Westermann, G., & Mareschal, D. 2014. From perceptual to language-mediated categorization. *Philosophical transactions of the Royal Society of London. Series B, Biological sciences*, 369(1634), 20120391.

Williams, G., Fabrizi, L., Meek, J., Jackson, D., Tracey, I., Robertson, N., Slater, R., & Fitzgerald, M. 2015. Functional magnetic resonance imaging can be used to explore tactile and nociceptive processing in the infant brain. *Acta Paediatrica*, 158–66.

Yong, M. H., & Ruffman, T. 2014. Emotional contagion: dogs and humans show a similar physiological response to human infant crying. *Behavioural Processes*, 108, 155–65.

Part II

5 Instrumental assessment of residual consciousness in DOCs

Carlo Cavaliere, Carol Di Perri, Steven Laureys, and Andrea Soddu

Introduction

Advances in emergency medicine, intensive care, and reanimation have significantly increased the number of patients that survive traumatic accidents, prolonged cardiac arrest, and severe brain injures of different etiologies (Laureys and Boly, 2008).

A considerable percentage of these patients will not fully recover primary functions over time and will suffer from different conditions of altered consciousness.

Disorders of consciousness (DOC) include a wide spectrum of clinical conditions (Bernat, 2006; Giacino et al., 2002; Laureys et al., 2004b; Plum and Posner, 1972; Schiff, 2006) encompassing states of unawareness, from vegetative state/unresponsive wakefulness syndrome (VS/UWS [Laureys et al., 2010]) to transient signs of consciousness (minimally conscious state [MCS] [Giacino et al., 2002]) and apparent unawareness, with full consciousness retained (locked-in syndrome [LiS] [Bauer et al., 1979]).

Although there is a uniform classification system based on defined behavioral criteria, the assessment of potential awareness in patients with DOC is still challenging, leading to delicate medical and ethical issues in contemporary neuroscience (Owen and Coleman, 2008; Farisco and Petrini, 2014).

Until now, the gold standard for the evaluation of the level of consciousness in patients with DOC is still the bedside clinical assessment (Boly et al., 2012), with a percentage of misdiagnosis up to 40 percent (Majerus et al., 2005; Schnakers et al., 2009). The use of specific standardized clinical scales, like the Coma Recovery Scale-Revised (CRS-R), have shown to significantly decrease this percentage of potential misdiagnosis, especially in the acute setting (Giacino et al., 2004) (American Congress of Rehabilitation Medicine et al., 2010). The Full Outline of UnResponsiveness (FOUR), instead, represents a valid alternative for LIS diagnosis and visual pursuit evaluation (Bruno et al., 2011).

Bedside clinical assessment evaluates patients' responsiveness. However, we should keep in mind that the absence of responsiveness does not necessary imply the absence of awareness (Monti et al., 2010). Indeed, the clinical assessment in brain-damaged patients remains difficult, subjective, and deeply affected by a patient's disabilities, such as paralysis or aphasia, cortical deafness or blindness, limited attentional capacities, or effects induced by pharmacological treatment (Giacino et al., 2004; Monti et al., 2010).

As such, the integration of neuroimaging techniques can overcome the subjectivity of the clinical assessment and might have an important role in improving our diagnosis in this challenging group of patients (Giacino et al., 2014). Indeed, these techniques can provide objective diagnostic and prognostic markers in patients with DOC, including morphological information regarding lesions site and extent (computed tomography [CT] and conventional structural Magnetic Resonance Imaging [MRI]), structural connectivity and white matter integrity (Diffusion Tensor Imaging [DTI]), brain metabolism (Magnetic Resonance Spectroscopy [MRS] and Positron Emission Tomography with Fluore-18 fluorodeoxyglucose tracer [18F-FDG-PET]), cortical hemodynamic response (functional Magnetic Resonance Imaging [fMRI] and 15O-H2O-PET), cortical fluctuations and response (Electroencephalography [EEG] and Transcranial Magnetic Stimulation [TMS]) (Tshibanda et al., 2010; Di Perri et al., 2014).

The aim of this chapter is to discuss the added value of these neuroimaging modalities in the diagnostic and prognostic assessment of patients with DOC.

Neuroimaging techniques: structural imaging

CT and MR techniques, due to their high spatial and contrast resolution, are considered among the most informative modalities for structural imaging.

Compared to MRI, CT scanning is the gold standard imaging technique in the acute setting, considering its high accessibility in the emergency, speed of acquisition, and sensitivity to hemorrhagic lesions (Giacino et al., 2014). However, the IMPACT score (International Mission for Prognosis and Analysis of Clinical Trials) (Steyerberg et al., 2008) derived from clinical, biochemical, and CT findings, have not shown a discriminative value for clinical decision making in patients (Lingsma et al., 2010; Maas et al., 2007).

Conversely, MRI, for its higher contrast resolution and multi-parametric acquisitions, represents the method of choice for visualizing brain damage extension and location in patients with chronic DOC, and monitoring cortical/ subcortical atrophy (Trivedi et al., 2007) and lesions evolution. Several studies have demonstrated the utility of MRI with conventional sequences (T1- and T2-weighted TSE, FLAIR, and Diffusion-Weighted Imaging [DWI]) in patients with traumatic brain injury (TBI) (Mannion et al., 2007; Skandsen et al., 2011; Yanagawa et al., 2009) and hypoxic/anoxic injury (HAI) following cardiac arrest (Topcuoglu et al., 2009; Jarnum et al., 2009).

For TBI patients, lesions in the corpus callosum, basal ganglia, and brain stem, especially if bilateral, were predictive of unfavorable outcome, also in acute setting (Firsching et al., 1998; Carpentier et al., 2006; Hoelper et al., 2000). Recently, also DWI has been standardized and included in the standard sequences for brain assessment, although not strictly a structural sequence. This approach, and the derived apparent diffusion coefficient (ADC), has been employed mainly for the evaluation of patients with HAI (Topcuoglu et al., 2009; Jarnum et al., 2009; Wijman et al., 2009), sometimes with controversial findings. Several authors (Wu et al., 2009) found a lower value of whole brain median ADC in poor-outcome patients; others (Wijman et al., 2009) suggested that the site and extent of brain lesions is more indicative than the raw decrease in the overall mean ADC value for outcome prediction. A third group (Choi et al., 2010), analyzing 39 HAI patients, reported an inverse correlation between cortex and basal ganglia lesions with the outcome, and identified ADC threshold for outcome prediction with 100 percent specificity.

Despite their encouraging results, these methods failed to explain why some patients with a clinical severe impairment have minimal brain lesions on conventional MRI examination, highlighting the lack of specificity and sensitivity of conventional MRI when used alone in the assessment of patients with DOC and suggesting the need of a multimodal approach with different advanced functional neuroimaging techniques (Figure 5.1).

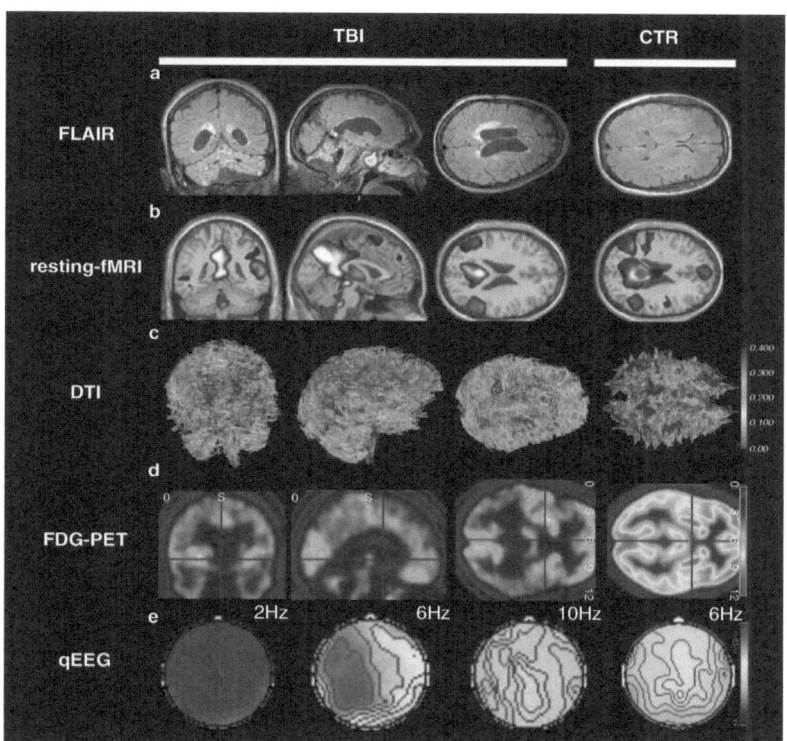

Figure 5.1 Multimodal neuroimaging analysis for a 30-year-old patient, 20 months after post-traumatic brain injury (TBI).

(a) Structural imaging in the three orthogonal planes. FLAIR sequence shows asymmetric dilatation of ventricles with peri-ventricular hyperintensity due to gliosis. (b) fMRI during resting state. Partial preservation of DMN, and in particular of the posterior components is shown. (c) DTI reconstruction of whole-brain tractography. Colors represent a scalar mode of FA values, demonstrating a global reduction of FA in TBI patients. (d) FDG-PET during resting state. A diffuse brain hypo-metabolism is shown. (e) High-density qEEG analysis with topographical distribution at a frequency of 2, 6, and 10 Hz. In the right column, a representative control image (CTR) for each technique.

Neuroimaging techniques: functional imaging

Key advances in our understanding of DOC have come from the use of functional neuroimaging, which gave us the possibility to objectively measure cognitive processing and complex functions, also in unresponsive patients (Di Perri et al., 2014).

Based on the imaging modality employed, functional neuroimaging can measure structural connectivity and white matter integrity (e.g., by use of DTI), brain metabolism (e.g., by use of MRS and 18F-FDG-PET), cortical hemodynamic response (e.g., by use of fMRI and 15O-H2O-PET), cortical electrical fluctuations and response (e.g., by use of EEG and TMS).

Depending on the acquisition paradigm, these techniques can measure brain functions during rest or active brain tasks, in response to passive external stimulation or active cognitive conditions.

Initially, functional neuroimaging studies in DOC have employed "block paradigms" with both passive tasks, to investigate residual functions in DOC patients, and active tasks, to further demonstrate command following. More recent approaches to study the networking supposedly supporting conscious processes consist in measuring spontaneous brain activity without the need for any stimulation or patient's compliance (Laureys et al., 2004a). This approach, called also resting-state fMRI, because it does not require patients' compliance, is particularly appealing in the study of severe brain injured unresponsive patients.

fMRI

fMRI is an MRI technique that measures the blood oxygenation level dependent (BOLD) changes that occur in response to neural activity. Three kinds of paradigm are commonly used to study brain activity: passive and active paradigms and resting state. fMRI activation studies with passive paradigm have recently replaced PET research with H20 tracer and sensory stimuli. These studies, using mainly auditory stimulation, have shown the presence of near normal high-level cortical activation in MCS patients compared to lower level activation in VS/UWS patients (Bekinschtein et al., 2005; Fernandez-Espejo et al., 2008; Coleman et al., 2009). Furthermore, a peculiar disconnection between primary sensory cortices and associative ones have been also reported in auditory (Laureys et al., 2000; Boly et al., 2005), somatosensory (Boly et al., 2005, 2008), or visual (Owen et al., 2002) stimulations. Moreover, it has been shown with passive fMRI paradigm that some VS/UWS patients, exhibiting brain activation comparable to that of MCS patients (i.e., atypical – not confined to primary sensory areas), have recovered signs of consciousness in long-term follow-up. For example, in a group of eight patients diagnosed as VS/UWS, brain activation to the sound of her/his own name spoken by a familiar voice elicited an atypical activation – suggestive of MCS – in two patients. Interestingly, those two patients subsequently recovered to MCS (Di et al., 2007). This suggested also a prognostic value for fMRI. However, we should keep in mind that the lack of activation to passive stimuli in DOC patients does not unequivocally coincide with the lack of awareness, as sensory deficits often present in this peculiar patient population might lead to false negatives.

Compared to passive external stimulation (e.g., auditory, somatosensory, and visual), active cognitive tasks have recently added new insights into the understanding of complex pathophysiology of individual DOC patients (Monti et al., 2010).

Recent fMRI studies using mental imagery tasks have demonstrated the ability of some patients to willfully modulate their brain activation, and in one case also to perform binary communication (Monti et al., 2010).

Only recently have we been able to study the spontaneous brain activity. Resting-state fMRI is a non-invasive technique used to investigate the spontaneous temporal coherence in BOLD fluctuations related to the amount of synchronized neural activity (i.e., functional connectivity) existing between distinct brain locations, in the absence of input or output tasks (Biswal et al., 1997). This technique has been increasingly used in the analysis of patients with DOC, mainly because it is not invasive and it surpasses the requirement for motor output or language comprehension. Among the several functional networks that have been detected so far (Beckmann et al., 2005), the default mode network (DMN) has been the most studied.

Various studies have shown a reduced connectivity of the DMN in disorders of consciousness which correlated with the level of consciousness (Vanhaudenhuyse et al., 2010; Tshibanda et al., 2010). However, to date, resting-state fMRI findings are not yet conclusive at single subject level and its exact functional role is still unknown.

Next to the findings of reduced connectivity in DMN, it has been recently shown pathological hyperconnectivity between the DMN and areas outside the DMN, such as the subcortical limbic system, including the orbitofrontal cortex and the insula. These findings suggest that the presence of hyper-connectivity patterns might also be informative of patients' brain function (Di Perri et al., 2013).

Furthermore, it has been found that the resting brain is characterized by a switch between the dominance of the DMN (linked to "internal" or self-awareness) and the bilateral frontoparietal network (linked to "external" or environmental awareness [Fox et al., 2005; Fransson, 2005]), i.e., when one is active the other is not and vice-versa. These spontaneous anticorrelated patterns have been shown to be related to mentation and behavioral status (Vanhaudenhuyse et al., 2011). In states of reduced consciousness, including a case of VS/UWS (Demertzi et al., 2015) a decreased anticorrelation between these two networks has been observed, suggesting the functional relevance of anticorrelated patterns to the phenomenological complexity of consciousness (Demertzi et al., 2013).

In conclusions, brain connectivity studies using resting-state fMRI have depicted a complex scenario of altered brain connectivity patterns characterized by pathologically increased and decreased connectivity patterns and anticorrelation alterations (Vanhaudenhuyse et al., 2010; Di Perri et al., 2013; Heine et al., 2012). Further studies will elucidate the functional meaning of these findings.

DTI

DTI is an advanced MR sequence (Basser et al., 1994), which can assess microstructural white matter integrity, also in absence of clear lesions to the structural imaging techniques (Arfanakis et al., 2002b; Hulkower et al., 2013).

It represents an extension of DWI, and it is basically focused on the estimation of water molecule diffusion along the direction of neuronal fibers tracts. After acquisition, different methods to reconstruct fiber orientation have been developed based on deterministic, probabilistic, or deconvolution algorithms (Cavaliere et al., 2014). Compared to other functional neuroimaging techniques applied to DOC, the principal advantage of DTI is that it allows evaluating structural connectivity also in sedated patients, not being influenced by sedatives or hypnotic drugs.

Up to now, numerous evidences have been published on DTI, mainly in TBI patients, highlighting the role of this technique as a valuable biomarker for the severity of tissue injury and a predictor for outcome (Huisman et al., 2004).

Limited studies on AHI (Gerdes et al., 2014; Luyt et al., 2012; Newcombe et al., 2010) have reported a significant reduction of fractional anisotropy values in predefined supratentorial regions (e.g., corpus callosum and thalamus), with no alterations of infratentorial regions (e.g., brain stem). The authors concluded that following AHI, FA values show a sensitivity of 94 percent and a specificity of 100 percent to predict unfavorable outcome (Luyt et al., 2012). A considerable amount of literature has instead focused on TBI patients and DTI alterations (Arfanakis et al., 2002a). Reduced values of fractional anisotropy have been reported in many regions of interest, including corona radiata, cortico-spinal tracts, cingulum, external capsule, corpus callosum, inferior fronto-occipital fasciculus, superior longitudinal fasciculus, and sagittal stratum (Kraus et al., 2007; Perlbarg et al., 2009; Edlow et al., 2013; Galanaud et al., 2012). These alterations were negatively correlated to clinical scores and outcome, especially for callosum alterations (Rutgers et al., 2008; Kraus et al., 2007; Newcombe et al., 2011). A peculiar damage of infratentorial structures has been identified in TBI compared to AHI patients (Newcombe et al., 2010). Few studies have reported a role for the ascending arousal reticular system in TBI patients (Edlow et al., 2013; McNab et al., 2013), showing a disruption of fibers projecting from brain stem to thalamic nuclei, and a partial preservation of thalami-cortical projections. Finally, a recent study has used DTI to discriminate between MCS and VS patients, demonstrating a power diagnostic accuracy for DTI of 95 percent (Fernandez-Espejo et al., 2011).

MRS

MRS is an advanced sequence of MR imaging, able to quantify in a specific region of interest the concentration of several metabolites (Cecil et al., 1998). 1H (proton) is the most commonly detected nucleus in MRS, although other nuclei can be used (e.g., 23Na or 31P). The main metabolites identified with proton

MRS are N-acetyl aspartate (NAA), an index of neuronal and axonal integrity; choline (Cho), linked to membrane turnover and integrity; creatine and phosphocreatine peak (Cr), a marker for metabolism of brain energy and tissue death; and lactate (La), an index for glycolysis and oxygen deficiency.

Also in the case of proton MRS, most studies have focused on TBI patients, and have often provided controversial results. These inconsistencies are due to heterogeneous samples, in terms of patient selection, delay from injury, voxel position, and acquisition protocols (Weiss et al., 2007).

Several authors reported a reduction of NAA value a few minutes after a TBI, with a minimum within 48 hours and a plateau that remains stable within the first month from the injury (Holshouser et al., 2006; Signoretti et al., 2002). For this reason, NAA concentration represents a good prognostic marker in the acute phase following injury, correlating with clinical assessment. Later, NAA value becomes too heterogeneous, limiting this biomarker to acute DOC.

Other studies (Ricci et al., 1997) have demonstrated that reduction of NAA/Cho ratio following TBI correlates with the unfavorable outcome of severe TBI patients.

However, NAA/Cr ratio reduction seems to be a better index of poor outcome than NAA/Cho ratio (Cecil et al., 1998). Reduced levels of NAA/Cr have been detected in gray and white matter of occipito-parietal (Ross et al., 1998; Friedman et al., 1999), frontal (Garnett et al., 2000), splenium of corpus callosum (Sinson et al., 2001), thalami (Uzan et al., 2003), and pons (Carpentier et al., 2006). Furthermore, La levels seem to increase following TBI (Marino et al., 2007) and to correlate with clinical outcome.

PET

PET is a nuclear medicine technique that produces functional brain images detecting the signal emitted by a radionuclide (tracer). PET quantification using a freely diffusible tissue tracer, like 15O-H2O, will indicate regional cerebral blood flow (rCBF), and therefore brain active areas. PET quantification with other tracers, for instance 18F-FDG, an analogue of glucose, will indicate tissue metabolic activity, that is regional glucose uptake.

As for fMRI with passive tasks, 15H-H2O PET studies using auditory and noxious stimulation (Laureys et al., 2002, 2000) have confirmed a disconnection between primary and associative sensory cortices in VS/UWS patients, with a partial preservation of fronto-parietal network in MCS (Laureys et al., 2004a). Similar findings have been observed for nociceptive stimuli, with a partial preservation of pain pathways in MCS and a limited activation to primary somatosensory areas in VS/UWS (Boly et al., 2008).

As far as concerns 18F-FDG PET during "resting state" conditions, DOC patients have shown a significant decrease of cerebral metabolic rate of glucose (about 42 and 55 percent of control values for VS/UWS and MCS, respectively) (Stender et al., 2015; Laureys et al., 2004b). These differences are most evident in the fronto-parietal regions (Laureys et al., 1999). Moreover, recovery of consciousness seems to be characterized by recovery of activity in this network, and between these regions and the thalamus (thalamo-cortical connections) (Laureys et al., 1999). Similar fronto-parietal hypometabolism has also been detected in other states of reduced consciousness, such as deep sleep (Maquet, 2000) and general anesthesia (Darkner and Sporring, 2013).

EEG

EEG is a common non-invasive technique used to measure brain electrical activity during rest condition or specific stimuli through multiple electrodes placed on the scalp. Compared to other neuroimaging techniques, EEG has a higher temporal resolution, is cheaper, requires no patient's compliance, and is less affected by movement artifacts or metal implants that can contraindicate MR techniques. In contrast, electrode failure, subjectivity of assessment, and duration of recording are common issues for this technology.

EEG includes a palette of different methods going over the standard assessment, like quantitative EEG (qEEG) analysis and event-related potentials (ERP).

Although with a limited spatial resolution that affects diagnostic power of this method (Guerit, 2007), standard EEG provides important prognostic information, already at bedside of DOC patients (Lehembre et al., 2012; Malinowska et al., 2013). A first visible effect is represented by the slowing of brain activity proportional to the severity of TBI/AHI injury with a non-linear transition from posterior alpha to diffuse theta or delta rhythm. Alpha/theta coma activity and burst suppression patterns are index of unfavorable prognosis (Kaplan et al., 1999). The EEG reaction to active stimulation (like auditory or noxious) is a sign of better prognosis.

To overcome subjectivity of standard EEG, qEEG is able to compute several parameters extracted from the EEG, like power spectra, connectivity values, or entropy, to score and classify different consciousness states (Pereda et al., 2005; Schiff et al., 2007; Thatcher, 2010). Several studies (Lehembre et al., 2012) have reported alterations in EEG power spectra of DOC patients, showing an increase of delta-power and a decrease of alpha-power for UWS/VS, when compared to MCS.

When coupling parameters derived from two electrodes, it is possible to estimate connectivity of underlying regions, an index disrupted in patients affected by DOC (Laureys, 2005). Other studies (Lehembre et al., 2012) confirm a significant reduction of connectivity in the theta and alpha bands for UWS/VS, when compared to MCS.

A third EEG approach, named as ERP, can be used to assess specific cortical function by measuring responses to repeated specific stimuli. Depending on the kind of stimulation, ERP can be divided in two categories: short latency or exogenous components, elicited by external stimulations, and cognitive or endogenous components, used to assess cognitive residual functions (passive paradigms) and consciousness (active paradigms) (Luck, 2005).

The absence of short latency ERP has been addressed as predictive of unfavorable outcome (Estraneo et al., 2013). Auditory ERP with processing of novelty or sound deviance can predict good outcome in DOC patients (Tzovara et al., 2013).

Similarly, auditory stimuli with emotional or auto referential valence can have prognostic value in these patients (Qin et al., 2008).

TMS

TMS is a non-invasive technique that activates specific cortical regions through the induction of brief and strong magnetic pulses generated by a dedicated coil applied on the scalp (Hallett, 2000). TMS does not require a scanner and it does not rely on the subject's ability to process sensory stimuli, to understand and follow instructions or to communicate. On the other hand, TMS stimulation can potentially induce epilepsy or stimulate damaged regions that cannot respond to the stimuli, biasing the assessment of consciousness.

Few studies have been published using TMS in DOC patients, and mainly coupling TMS with other electro-physiological techniques. Decreased motor cortex excitability has been detected with TMS of the motor cortex coupled with EMG; this decrease has been correlated to the level of consciousness (Lapitskaya et al., 2013). The majority of studies, instead, coupled TMS with EEG in order to assess brain connectivity (Boly et al., 2012). This technique does not require active compliance of DOC patients (Massimini et al., 2009; Rosanova et al., 2012) and, more importantly, can give information at the single subject level.

EEG-TMS can measure brain complexity combining the use of TMS to stimulate brain circuitry and high density EEG to quantify the effects of this perturbation on brain circuitry (Casali et al., 2013; Massimini et al., 2005).

Based on the level of consciousness, TMS activation will induce either cortical interaction due to a preservation of brain function, or loss of information and/or integration. For example, frontal TMS activation elicited no response or triggered a simple, local EEG response in UWS/VS patients. Conversely, in MCS, the same stimulus determined a complex and multisite activation, similar to LIS and healthy subjects (Rosanova et al., 2012). The local slow wave response recorded in VS patients indicated a disruption of effective intracortical connectivity (Friston, 2011). Moreover, these data highlighted the ability of TMS-EEG to better discriminate between different level of consciousness (Ragazzoni et al., 2013), and from a longitudinal point of view to evaluate the recovery of consciousness in DOC patients (Rosanova et al., 2012).

Recently, an index has been created to quantify these effects: the perturbational complexity index (PCI) (Casali et al., 2013). This promising TMS-EEG metric allows to discriminate in single subjects the level of consciousness: below 0.31 for unconsciousness, above 0.51 for healthy consciousness and in the between for MCS (Casali et al., 2013).

Conclusions

The assessment of patients with disorders of consciousness remains a big challenge. The bedside assessment is nowadays still offering the gold standard for the evaluation of level of consciousness. However, it is prone to misdiagnosis as it is subjective and depends entirely on patient's responsiveness.

In this context neuroimaging techniques have an important role in integrating the behavioral evaluation as they can give objectives information without relying solely on patient's responsiveness. We have here reviewed how the main neuroimaging techniques (PET, structural MRI, DTI fMRI, EEG, TMS, and TMS-EEG) can provide us with important insights about brain functioning in DOC patients and can help in the diagnosis and prognosis of DOC patients.

We expect that the ongoing development of the imaging techniques and analyses will improve our understanding of consciousness and our clinical approach to patients with DOC.

Our future challenge is to make this technique a valuable tool for radiologists and neurologists in the routinely clinical assessment. Indeed, apart from structural imaging, the use of these techniques remains still very limited. A further effort should be done to put clinicians in the condition to interpret imaging analysis and to combine them with the behavioral evaluations. We believe that specialized centers in imaging analysis might help by developing and offering services to the clinical facilities and more effort needs to be focused on the creation of automatized software, which could speed up the analysis of an exponential increasing amount of imaging data.

References

American Congress of Rehabilitation Medicine, B. I.-I. S. I. G. D. O. C. T. F., Seel, R. T., Sherer, M., Whyte, J., Katz, D. I., Giacino, J. T., Rosenbaum, A. M., Hammond, F. M., Kalmar, K., Pape, T. L., Zafonte, R., Biester, R. C., Kaelin, D., Kean, J., & Zasler, N. 2010. Assessment scales for disorders of consciousness: Evidence-based recommendations for clinical practice and research. *Arch Phys Med Rehabil*, 91, 1795–813.

Arfanakis, K., Cordes, D., Haughton, V. M., Carew, J. D., & Meyerand, M. E. 2002a. Independent component analysis applied to diffusion tensor MRI. *Magn Reson Med*, 47, 354–63.

Arfanakis, K., Haughton, V. M., Carew, J. D., Rogers, B. P., Dempsey, R. J., & Meyerand, M. E. 2002b. Diffusion tensor MR imaging in diffuse axonal injury. *AJNR Am J Neuroradiol*, 23, 794–802.

Basser, P. J., Mattiello, J., & Lebihan, D. 1994. MR diffusion tensor spectroscopy and imaging. *Biophys J*, 66, 259–67.

Bauer, G., Gerstenbrand, F., & Rumpl, E. 1979. Varieties of the locked-in syndrome. *J Neurol*, 221, 77–91.

Beckmann, C. F., Deluca, M., Devlin, J. T., & Smith, S. M. 2005. Investigations into resting-state connectivity using independent component analysis. *Philos Trans R Soc Lond B Biol Sci*, 360, 1001–13.

Bekinschtein, T., Tiberti, C., Niklison, J., Tamashiro, M., Ron, M., Carpintiero, S., Villarreal, M., Forcato, C., Leiguarda, R., & Manes, F. 2005. Assessing level of consciousness and cognitive changes from vegetative state to full recovery. *Neuropsychol Rehabil*, 15, 307–22.

Bernat, J. L. 2006. Chronic disorders of consciousness. *Lancet*, 367, 1181–92.

Biswal, B. B., Van Kylen, J., & Hyde, J. S. 1997. Simultaneous assessment of flow and BOLD signals in resting-state functional connectivity maps. *NMR Biomed*, 10, 165–70.

Boly, M., Faymonville, M. E., Peigneux, P., Lambermont, B., Damas, F., Luxen, A., Lamy, M., Moonen, G., Maquet, P., & Laureys, S. 2005. Cerebral processing of auditory and noxious stimuli in severely brain injured patients: Differences between VS and MCS. *Neuropsychol Rehabil*, 15, 283–9.

Boly, M., Faymonville, M. E., Schnakers, C., Peigneux, P., Lambermont, B., Phillips, C., Lancellotti, P., Luxen, A., Lamy, M., Moonen, G., Maquet, P., & Laureys, S. 2008. Perception of pain in the minimally conscious state with PET activation: An observational study. *Lancet Neurol*, 7, 1013–20.

Boly, M., Massimini, M., Garrido, M. I., Gosseries, O., Noirhomme, Q., Laureys, S., & Soddu, A. 2012. Brain connectivity in disorders of consciousness. *Brain Connect*, 2, 1–10.

Bruno, M. A., Ledoux, D., Lambermont, B., Damas, F., Schnakers, C., Vanhaudenhuyse, A., Gosseries, O., & Laureys, S. 2011. Comparison of the full outline of unresponsiveness and Glasgow Liege Scale/Glasgow Coma Scale in an intensive care unit population. *Neurocrit Care*, 15, 447–53.

Carpentier, A., Galanaud, D., Puybasset, L., Muller, J. C., Lescot, T., Boch, A. L., Riedl, V., Cornu, P., Coriat, P., Dormont, D., & Van Effenterre, R. 2006. Early morphologic and spectroscopic magnetic resonance in severe traumatic brain injuries can detect "invisible brain stem damage" and predict "vegetative states." *J Neurotrauma*, 23, 674–85.

Casali, A. G., Gosseries, O., Rosanova, M., Boly, M., Sarasso, S., Casali, K. R., Casarotto, S., Bruno, M. A., Laureys, S., Tononi, G., & Massimini, M. 2013. A theoretically based index of consciousness independent of sensory processing and behavior. *Sci Transl Med*, 5, 198ra105.

Cavaliere, C., Aiello, M., Di Perri, C., Fernandez-Espejo, D., Owen, A. M., & Soddu, A. 2014. Diffusion tensor imaging and white matter abnormalities in patients with disorders of consciousness. *Front Hum Neurosci*, 8, 1028.

Cecil, K. M., Hills, E. C., Sandel, M. E., Smith, D. H., Mcintosh, T. K., Mannon, L. J., Sinson, G. P., Bagley, L. J., Grossman, R. I., & Lenkinski, R. E. 1998. Proton magnetic resonance spectroscopy for detection of axonal injury in the splenium of the corpus callosum of brain-injured patients. *J Neurosurg*, 88, 795–801.

Choi, S. P., Park, K. N., Park, H. K., Kim, J. Y., Youn, C. S., Ahn, K. J., & Yim, H. W. 2010. Diffusion-weighted magnetic resonance imaging for predicting the clinical outcome of comatose survivors after cardiac arrest: A cohort study. *Crit Care*, 14, R17.

Coleman, M. R., Davis, M. H., Rodd, J. M., Robson, T., Ali, A., Owen, A. M., & Pickard, J. D. 2009. Towards the routine use of brain imaging to aid the clinical diagnosis of disorders of consciousness. *Brain*, 132, 2541–52.

Darkner, S., & Sporring, J. 2013. Locally orderless registration. *IEEE Trans Pattern Anal Mach Intell*, 35, 1437–50.

Demertzi, A., Antonopoulos, G., Heine, L., Voss, H. U., Crone, J. S., De Los Angeles, C., Bahri, M. A., Di Perri, C., Vanhaudenhuyse, A., Charland-Verville, V., Kronbichler, M., Trinka, E., Phillips, C., Gomez, F., Tshibanda, L., Soddu, A., Schiff, N. D., Whitfield-Gabrieli, S., & Laureys, S. 2015. Intrinsic functional connectivity differentiates minimally conscious from unresponsive patients. *Brain*, 138, 2619–31.

Demertzi, A., Vanhaudenhuyse, A., Bredart, S., Heine, L., Di Perri, C., & Laureys, S. 2013. Looking for the self in pathological unconsciousness. *Front Hum Neurosci*, 7, 538.

Di, H. B., Yu, S. M., Weng, X. C., Laureys, S., Yu, D., Li, J. Q., Qin, P. M., Zhu, Y. H., Zhang, S. Z., & Chen, Y. Z. 2007. Cerebral response to patient's own name in the vegetative and minimally conscious states. *Neurology*, 68, 895–9.

Di Perri, C., Bastianello, S., Bartsch, A. J., Pistarini, C., Maggioni, G., Magrassi, L., Imberti, R., Pichiecchio, A., Vitali, P., Laureys, S., & Di Salle, F. 2013. Limbic hyperconnectivity in the vegetative state. *Neurology*, 81, 1417–24.

Di Perri, C., Thibaut, A., Heine, L., Soddu, A., Demertzi, A., & Laureys, S. 2014. Measuring consciousness in coma and related states. *World J Radiol*, 6, 589–97.

Edlow, B. L., Haynes, R. L., Takahashi, E., Klein, J. P., Cummings, P., Benner, T., Greer, D. M., Greenberg, S. M., Wu, O., Kinney, H. C., & Folkerth, R. D. 2013. Disconnection of the ascending arousal system in traumatic coma. *J Neuropathol Exp Neurol*, 72, 505–23.

Estraneo, A., Moretta, P., Loreto, V., Lanzillo, B., Cozzolino, A., Saltalamacchia, A., Lullo, F., Santoro, L., & Trojano, L. 2013. Predictors of recovery of responsiveness in prolonged anoxic vegetative state. *Neurology*, 80, 464–70.

Farisco, M., & Petrini, C. 2014. Misdiagnosis as an ethical and scientific challenge. *Ann Ist Super Sanita*, 50, 229–33.

Fernandez-Espejo, D., Bekinschtein, T., Monti, M. M., Pickard, J. D., Junque, C., Coleman, M. R., & Owen, A. M. 2011. Diffusion weighted imaging distinguishes the vegetative state from the minimally conscious state. *Neuroimage*, 54, 103–12.

Fernandez-Espejo, D., Junque, C., Vendrell, P., Bernabeu, M., Roig, T., Bargallo, N., & Mercader, J. M. 2008. Cerebral response to speech in vegetative and minimally conscious states after traumatic brain injury. *Brain Inj*, 22, 882–90.

Firsching, R., Woischneck, D., Diedrich, M., Klein, S., Ruckert, A., Wittig, H., & Dohring, W. 1998. Early magnetic resonance imaging of brainstem lesions after severe head injury. *J Neurosurg*, 89, 707–12.

Fox, M. D., Snyder, A. Z., Vincent, J. L., Corbetta, M., Van Essen, D. C., & Raichle, M. E. 2005. The human brain is intrinsically organized into dynamic, anticorrelated functional networks. *Proc Natl Acad Sci U S A*, 102, 9673–8.

Fransson, P. 2005. Spontaneous low-frequency BOLD signal fluctuations: An fMRI investigation of the resting-state default mode of brain function hypothesis. *Hum Brain Mapp*, 26, 15–29.

Friedman, S. D., Brooks, W. M., Jung, R. E., Chiulli, S. J., Sloan, J. H., Montoya, B. T., Hart, B. L., & Yeo, R. A. 1999. Quantitative proton MRS predicts outcome after traumatic brain injury. *Neurology*, 52, 1384–91.

Friston, K. J. 2011. Functional and effective connectivity: A review. *Brain Connect*, 1, 13–36.

Galanaud, D., Perlbarg, V., Gupta, R., Stevens, R. D., Sanchez, P., Tollard, E., De Champfleur, N. M., Dinkel, J., Faivre, S., Soto-Ares, G., Veber, B., Cottenceau, V., Masson, F., Tourdias, T., Andre, E., Audibert, G., Schmitt, E., Ibarrola, D., Dailler, F., Vanhaudenhuyse, A., Tshibanda, L., Payen, J. F., Le Bas, J. F., Krainik, A., Bruder, N., Girard, N., Laureys, S., Benali, H., Puybasset, L., Neuro Imaging For Coma, E., & Recovery, C. 2012. Assessment of white matter injury and outcome in severe brain trauma: A prospective multicenter cohort. *Anesthesiology*, 117, 1300–10.

Garnett, M. R., Blamire, A. M., Corkill, R. G., Cadoux-Hudson, T. A., Rajagopalan, B., & Styles, P. 2000. Early proton magnetic resonance spectroscopy in normal-appearing brain correlates with outcome in patients following traumatic brain injury. *Brain*, 123(Pt 10), 2046–54.

Gerdes, J. S., Walther, E. U., Jaganjac, S., Makrigeorgi-Butera, M., Meuth, S. G., & Deppe, M. 2014. Early detection of widespread progressive brain injury after cardiac arrest: A single case DTI and post-mortem histology study. *PLoS One*, 9, e92103.

Giacino, J. T., Ashwal, S., Childs, N., Cranford, R., Jennett, B., Katz, D. I., Kelly, J. P., Rosenberg, J. H., Whyte, J., Zafonte, R. D., & Zasler, N. D. 2002. The minimally conscious state: Definition and diagnostic criteria. *Neurology*, 58, 349–53.

Giacino, J. T., Fins, J. J., Laureys, S., & Schiff, N. D. 2014. Disorders of consciousness after acquired brain injury: The state of the science. *Nat Rev Neurol*, 10, 99–114.

Giacino, J. T., Kalmar, K., & Whyte, J. 2004. The JFK Coma Recovery Scale-Revised: measurement characteristics and diagnostic utility. *Arch Phys Med Rehabil*, 85, 2020–9.

Guerit, J. M. 2007. Electroencephalography: The worst traditionally recommended tool for brain death confirmation. *Intensive Care Med*, 33, 9–10.

Hallett, M. 2000. Transcranial magnetic stimulation and the human brain. *Nature*, 406, 147–50.

Heine, L., Soddu, A., Gomez, F., Vanhaudenhuyse, A., Tshibanda, L., Thonnard, M., Charland-Verville, V., Kirsch, M., Laureys, S., & Demertzi, A. 2012. Resting state networks and consciousness: Alterations of multiple resting state network connectivity in physiological, pharmacological, and pathological consciousness states. *Front Psychol*, 3, 295.

Hoelper, B. M., Soldner, F., Chone, L., & Wallenfang, T. 2000. Effect of intracerebral lesions detected in early MRI on outcome after acute brain injury. *Acta Neurochir Suppl*, 76, 265–7.

Holshouser, B. A., Tong, K. A., Ashwal, S., Oyoyo, U., Ghamsary, M., Saunders, D., & Shutter, L. 2006. Prospective longitudinal proton magnetic resonance spectroscopic imaging in adult traumatic brain injury. *J Magn Reson Imaging*, 24, 33–40.

Huisman, T. A., Schwamm, L. H., Schaefer, P. W., Koroshetz, W. J., Shetty-Alva, N., Ozsunar, Y., Wu, O., & Sorensen, A. G. 2004. Diffusion tensor imaging as potential biomarker of white matter injury in diffuse axonal injury. *AJNR Am J Neuroradiol*, 25, 370–6.

Hulkower, M. B., Poliak, D. B., Rosenbaum, S. B., Zimmerman, M. E., & Lipton, M. L. 2013. A decade of DTI in traumatic brain injury: 10 years and 100 articles later. *AJNR Am J Neuroradiol*, 34, 2064–74.

Jarnum, H., Knutsson, L., Rundgren, M., Siemund, R., Englund, E., Friberg, H., & Larsson, E. M. 2009. Diffusion and perfusion MRI of the brain in comatose patients treated with mild hypothermia after cardiac arrest: A prospective observational study. *Resuscitation*, 80, 425–30.

Kaplan, P. W., Genoud, D., Ho, T. W., & Jallon, P. 1999. Etiology, neurologic correlations, and prognosis in alpha coma. *Clin Neurophysiol*, 110, 205–13.

Kraus, M. F., Susmaras, T., Caughlin, B. P., Walker, C. J., Sweeney, J. A., & Little, D. M. 2007. White matter integrity and cognition in chronic traumatic brain injury: A diffusion tensor imaging study. *Brain*, 130, 2508–19.

Lapitskaya, N., Gosseries, O., De Pasqua, V., Pedersen, A. R., Nielsen, J. F., De Noordhout, A. M., & Laureys, S. 2013. Abnormal corticospinal excitability in patients with disorders of consciousness. *Brain Stimul*, 6, 590–7.

Laureys, S. 2005. The neural correlate of (un)awareness: Lessons from the vegetative state. *Trends Cogn Sci*, 9, 556–9.

Laureys, S., & Boly, M. 2008. The changing spectrum of coma. *Nat Clin Pract Neurol*, 4, 544–6.

Laureys, S., Celesia, G. G., Cohadon, F., Lavrijsen, J., Leon-Carrion, J., Sannita, W. G., Sazbon, L., Schmutzhard, E., Von Wild, K. R., Zeman, A., Dolce, G., & European Task Force On Disorders Of, C. 2010. Unresponsive wakefulness syndrome: A new name for the vegetative state or apallic syndrome. *BMC Med*, 8, 68.

Laureys, S., Faymonville, M. E., Degueldre, C., Fiore, G. D., Damas, P., Lambermont, B., Janssens, N., Aerts, J., Franck, G., Luxen, A., Moonen, G., Lamy, M., & Maquet, P. 2000. Auditory processing in the vegetative state. *Brain*, 123 (Pt 8), 1589–601.

Laureys, S., Faymonville, M. E., Peigneux, P., Damas, P., Lambermont, B., Del Fiore, G., Degueldre, C., Aerts, J., Luxen, A., Franck, G., Lamy, M., Moonen, G., & Maquet, P. 2002. Cortical processing of noxious somatosensory stimuli in the persistent vegetative state. *Neuroimage*, 17, 732–41.

Laureys, S., Lemaire, C., Maquet, P., Phillips, C., & Franck, G. 1999. Cerebral metabolism during vegetative state and after recovery to consciousness. *J Neurol Neurosurg Psychiatry*, 67, 121.

Laureys, S., Owen, A. M., & Schiff, N. D. 2004a. Brain function in coma, vegetative state, and related disorders. *Lancet Neurol*, 3, 537–46.

Laureys, S., Perrin, F., Faymonville, M. E., Schnakers, C., Boly, M., Bartsch, V., Majerus, S., Moonen, G., & Maquet, P. 2004b. Cerebral processing in the minimally conscious state. *Neurology*, 63, 916–8.

Lehembre, R., Gosseries, O., Lugo, Z., Jedidi, Z., Chatelle, C., Sadzot, B., Laureys, S., & Noirhomme, Q. 2012. Electrophysiological investigations of brain function in coma, vegetative, and minimally conscious patients. *Arch Ital Biol*, 150, 122–39.

Lingsma, H. F., Roozenbeek, B., Steyerberg, E. W., Murray, G. D., & Maas, A. I. 2010. Early prognosis in traumatic brain injury: From prophecies to predictions. *Lancet Neurol*, 9, 543–54.

Luck, S. 2005. An introduction to the event-related potentials and their neural origins. In Luck, S. (ed.), *An introduction to the Event-Related Potential Technique*. Cambridge, MA: The MIT Press.

Luyt, C. E., Galanaud, D., Perlbarg, V., Vanhaudenhuyse, A., Stevens, R. D., Gupta, R., Besancenot, H., Krainik, A., Audibert, G., Combes, A., Chastre, J., Benali, H., Laureys, S., Puybasset, L., Neuro Imaging For Coma, E., & Recovery, C. 2012. Diffusion tensor imaging to predict long-term outcome after cardiac arrest: A bicentric pilot study. *Anesthesiology*, 117, 1311–21.

Maas, A. I., Steyerberg, E. W., Butcher, I., Dammers, R., Lu, J., Marmarou, A., Mushkudiani, N. A., Mchugh, G. S., & Murray, G. D. 2007. Prognostic value of computerized tomography scan characteristics in traumatic brain injury: Results from the IMPACT study. *J Neurotrauma*, 24, 303–14.

Majerus, S., Gill-Thwaites, H., Andrews, K., & Laureys, S. 2005. Behavioral evaluation of consciousness in severe brain damage. *Prog Brain Res*, 150, 397–413.

Malinowska, U., Chatelle, C., Bruno, M. A., Noirhomme, Q., Laureys, S., & Durka, P. J. 2013. Electroencephalographic profiles for differentiation of disorders of consciousness. *Biomed Eng Online*, 12, 109.

Mannion, R. J., Cross, J., Bradley, P., Coles, J. P., Chatfield, D., Carpenter, A., Pickard, J. D., Menon, D. K., & Hutchinson, P. J. 2007. Mechanism-based MRI classification of traumatic brain stem injury and its relationship to outcome. *J Neurotrauma*, 24, 128–35.

Maquet, P. 2000. Functional neuroimaging of normal human sleep by positron emission tomography. *J Sleep Res*, 9, 207–31.

Marino, S., Zei, E., Battaglini, M., Vittori, C., Buscalferri, A., Bramanti, P., Federico, A., & De Stefano, N. 2007. Acute metabolic brain changes following traumatic brain injury and their relevance to clinical severity and outcome. *J Neurol Neurosurg Psychiatry*, 78, 501–7.

Massimini, M., Boly, M., Casali, A., Rosanova, M., & Tononi, G. 2009. A perturbational approach for evaluating the brain's capacity for consciousness. *Prog Brain Res*, 177, 201–14.

Massimini, M., Ferrarelli, F., Huber, R., Esser, S. K., Singh, H., & Tononi, G. 2005. Breakdown of cortical effective connectivity during sleep. *Science*, 309, 2228–32.

Mcnab, J. A., Edlow, B. L., Witzel, T., Huang, S. Y., Bhat, H., Heberlein, K., Feiweier, T., Liu, K., Keil, B., Cohen-Adad, J., Tisdall, M. D., Folkerth, R. D., Kinney, H. C., & Wald, L. L. 2013. The Human Connectome Project and beyond: Initial applications of 300 mT/m gradients. *Neuroimage*, 80, 234–45.

Monti, M. M., Vanhaudenhuyse, A., Coleman, M. R., Boly, M., Pickard, J. D., Tshibanda, L., Owen, A. M., & Laureys, S. 2010. Willful modulation of brain activity in disorders of consciousness. *N Engl J Med*, 362, 579–89.

Newcombe, V., Chatfield, D., Outtrim, J., Vowler, S., Manktelow, A., Cross, J., Scoffings, D., Coleman, M., Hutchinson, P., Coles, J., Carpenter, T. A., Pickard, J., Williams, G., & Menon, D. 2011. Mapping traumatic axonal injury using diffusion tensor imaging: Correlations with functional outcome. *PLoS One*, 6, e19214.

Newcombe, V. F., Williams, G. B., Scoffings, D., Cross, J., Carpenter, T. A., Pickard, J. D., & Menon, D. K. 2010. Aetiological differences in neuroanatomy of the vegetative state: Insights from diffusion tensor imaging and functional implications. *J Neurol Neurosurg Psychiatry*, 81, 552–61.

Owen, A. M., & Coleman, M. R. 2008. Detecting awareness in the vegetative state. *Ann N Y Acad Sci*, 1129, 130–8.

Owen, A. M., Menon, D. K., Johnsrude, I. S., Bor, D., Scott, S. K., Manly, T., Williams, E. J., Mummery, C., & Pickard, J. D. 2002. Detecting residual cognitive function in persistent vegetative state. *Neurocase*, 8, 394–403.

Pereda, E., Quiroga, R. Q., & Bhattacharya, J. 2005. Nonlinear multivariate analysis of neurophysiological signals. *Prog Neurobiol*, 77, 1–37.

Perlbarg, V., Puybasset, L., Tollard, E., Lehericy, S., Benali, H., & Galanaud, D. 2009. Relation between brain lesion location and clinical outcome in patients with severe traumatic brain injury: A diffusion tensor imaging study using voxel-based approaches. *Hum Brain Mapp*, 30, 3924–33.

Plum, F., & Posner, J. B. 1972. The diagnosis of stupor and coma. *Contemp Neurol Ser*, 10, 1–286.

Qin, P., Di, H., Yan, X., Yu, S., Yu, D., Laureys, S., & Weng, X. 2008. Mismatch negativity to the patient's own name in chronic disorders of consciousness. *Neurosci Lett*, 448, 24–8.

Ragazzoni, A., Pirulli, C., Veniero, D., Feurra, M., Cincotta, M., Giovannelli, F., Chiaramonti, R., Lino, M., Rossi, S., & Miniussi, C. 2013. Vegetative versus minimally conscious states: A study using TMS-EEG, sensory and event-related potentials. *PLoS One*, 8, e57069.

Ricci, B., Barbarella, G., Musi, P., Boldrini, P., Trevisan, C., & Basaglia, N. 1997. Localised proton MR spectroscopy of brain metabolism changes in vegetative patients. *Neuroradiology*, 39, 313–19.

Rosanova, M., Gosseries, O., Casarotto, S., Boly, M., Casali, A. G., Bruno, M. A., Mariotti, M., Boveroux, P., Tononi, G., Laureys, S., & Massimini, M. 2012. Recovery of cortical effective connectivity and recovery of consciousness in vegetative patients. *Brain*, 135, 1308–20.

Ross, B. D., Ernst, T., Kreis, R., Haseler, L. J., Bayer, S., Danielsen, E., Bluml, S., Shonk, T., Mandigo, J. C., Caton, W., Clark, C., Jensen, S. W., Lehman, N. L., Arcinue, E., Pudenz, R., & Shelden, C. H. 1998. 1H MRS in acute traumatic brain injury. *J Magn Reson Imaging*, 8, 829–40.

Rutgers, D. R., Fillard, P., Paradot, G., Tadie, M., Lasjaunias, P., & Ducreux, D. 2008. Diffusion tensor imaging characteristics of the corpus callosum in mild, moderate, and severe traumatic brain injury. *AJNR Am J Neuroradiol*, 29, 1730–5.

Schiff, N. D. 2006. Multimodal neuroimaging approaches to disorders of consciousness. *J Head Trauma Rehabil*, 21, 388–97.

Schiff, N. D., Giacino, J. T., Kalmar, K., Victor, J. D., Baker, K., Gerber, M., Fritz, B., Eisenberg, B., Biondi, T., O'connor, J., Kobylarz, E. J., Farris, S., Machado, A., Mccagg, C., Plum, F., Fins, J. J., & Rezai, A. R. 2007. Behavioural improvements with thalamic stimulation after severe traumatic brain injury. *Nature*, 448, 600–3.

Schnakers, C., Vanhaudenhuyse, A., Giacino, J., Ventura, M., Boly, M., Majerus, S., Moonen, G., & Laureys, S. 2009. Diagnostic accuracy of the vegetative and minimally conscious state: Clinical consensus versus standardized neurobehavioral assessment. *BMC Neurol*, 9, 35.

Signoretti, S., Marmarou, A., Fatouros, P., Hoyle, R., Beaumont, A., Sawauchi, S., Bullock, R., & Young, H. 2002. Application of chemical shift imaging for measurement of NAA in head injured patients. *Acta Neurochir Suppl*, 81, 373–5.

Sinson, G., Bagley, L. J., Cecil, K. M., Torchia, M., Mcgowan, J. C., Lenkinski, R. E., Mcintosh, T. K., & Grossman, R. I. 2001. Magnetization transfer imaging and proton MR spectroscopy in the evaluation of axonal injury: Correlation with clinical outcome after traumatic brain injury. *AJNR Am J Neuroradiol*, 22, 143–51.

Skandsen, T., Kvistad, K. A., Solheim, O., Lydersen, S., Strand, I. H., & Vik, A. 2011. Prognostic value of magnetic resonance imaging in moderate and severe head injury: A prospective study of early MRI findings and one-year outcome. *J Neurotrauma*, 28, 691–9.

Stender, J., Kupers, R., Rodell, A., Thibaut, A., Chatelle, C., Bruno, M. A., Gejl, M., Bernard, C., Hustinx, R., Laureys, S., & Gjedde, A. 2015. Quantitative rates of brain glucose metabolism distinguish minimally conscious from vegetative state patients. *J Cereb Blood Flow Metab*, 35, 58–65.

Steyerberg, E. W., Mushkudiani, N., Perel, P., Butcher, I., Lu, J., Mchugh, G. S., Murray, G. D., Marmarou, A., Roberts, I., Habbema, J. D., & Maas, A. I. 2008. Predicting outcome after traumatic brain injury: Development and international validation of prognostic scores based on admission characteristics. *PLoS Med*, 5, e165; discussion e165.

Thatcher, R. 2010. Validity and reliability of quantitative Electroencephalography (qEEG). *Journal of Neurotherapy*, 14, 122–52.

Topcuoglu, M. A., Oguz, K. K., Buyukserbetci, G., & Bulut, E. 2009. Prognostic value of magnetic resonance imaging in post-resuscitation encephalopathy. *Intern Med*, 48, 1635–45.

Trivedi, M. A., Ward, M. A., Hess, T. M., Gale, S. D., Dempsey, R. J., Rowley, H. A., & Johnson, S. C. 2007. Longitudinal changes in global brain volume between 79 and 409 days after traumatic brain injury: Relationship with duration of coma. *J Neurotrauma*, 24, 766–71.

Tshibanda, L., Vanhaudenhuyse, A., Boly, M., Soddu, A., Bruno, M. A., Moonen, G., Laureys, S., & Noirhomme, Q. 2010. Neuroimaging after coma. *Neuroradiology*, 52, 15–24.

Tzovara, A., Rossetti, A. O., Spierer, L., Grivel, J., Murray, M. M., Oddo, M., & De Lucia, M. 2013. Progression of auditory discrimination based on neural decoding predicts awakening from coma. *Brain*, 136, 81–9.

Uzan, M., Albayram, S., Dashti, S. G., Aydin, S., Hanci, M., & Kuday, C. 2003. Thalamic proton magnetic resonance spectroscopy in vegetative state induced by traumatic brain injury. *J Neurol Neurosurg Psychiatry*, 74, 33–8.

Vanhaudenhuyse, A., Demertzi, A., Schabus, M., Noirhomme, Q., Bredart, S., Boly, M., Phillips, C., Soddu, A., Luxen, A., Moonen, G., & Laureys, S. 2011. Two distinct neuronal networks mediate the awareness of environment and of self. *J Cogn Neurosci*, 23, 570–8.

Vanhaudenhuyse, A., Noirhomme, Q., Tshibanda, L. J., Bruno, M. A., Boveroux, P., Schnakers, C., Soddu, A., Perlbarg, V., Ledoux, D., Brichant, J. F., Moonen, G., Maquet, P., Greicius, M. D., Laureys, S., & Boly, M. 2010. Default network connectivity reflects the level of consciousness in non-communicative brain-damaged patients. *Brain*, 133, 161–71.

Weiss, N., Galanaud, D., Carpentier, A., Naccache, L., & Puybasset, L. 2007. Clinical review: Prognostic value of magnetic resonance imaging in acute brain injury and coma. *Crit Care*, 11, 230.

Wijman, C. A., Mlynash, M., Caulfield, A. F., Hsia, A. W., Eyngorn, I., Bammer, R., Fischbein, N., Albers, G. W., & Moseley, M. 2009. Prognostic value of brain diffusion-weighted imaging after cardiac arrest. *Ann Neurol*, 65, 394–402.

Wu, O., Sorensen, A. G., Benner, T., Singhal, A. B., Furie, K. L., & Greer, D. M. 2009. Comatose patients with cardiac arrest: Predicting clinical outcome with diffusion-weighted MR imaging. *Radiology*, 252, 173–81.

Yanagawa, Y., Sakamoto, T., Takasu, A., & Okada, Y. 2009. Relationship between maximum intracranial pressure and traumatic lesions detected by T2*-weighted imaging in diffuse axonal injury. *J Trauma*, 66, 162–5.

6 Neurotechnological communication with patients with disorders of consciousness

Damien Lesenfants, Camille Chatelle, Jad Saab, Steven Laureys, and Quentin Noirhomme

Abbreviations

BCI	Brain-Computer Interface
CRS-R	Coma Recovery Scale-Revised
DOC	Disorders Of Consciousness
EEG	ElectroEncephaloGraphy
fMRI	functional Magnetic Resonance Imaging
LIS	Locked-In Syndrome
MCS	Minimally Conscious State
SSVEP	Steady-State Visually Evoked Potential
VS/UWS	Vegetative State/Unresponsive Wakefulness Syndrome

Beyond motor control: brain-computer interfaces

A brain-computer interface (BCI) is a system enabling a person (i.e., the user) to communicate with the external world without using traditional pathways such as peripheral nerves and muscles (see Figure 6.1) (Wolpaw et al., 2002). BCIs are also known as brain-machine interfaces or neuroprostheses. While the former terminology is used in the electroencephalography (EEG) and functional Magnetic Resonance Imaging (fMRI) brain interface communities, the latter is commonly used in the invasive recording brain interface community. Generally, a BCI aims to translate patterns of brain activity into computer commands and can be divided into four main components (Figure 6.1): signal acquisition, feature extraction, classification, and feedback/command. In this chapter, we will use the term "patient" to refer to an individual with motor and/or cognitive disabilities following a brain-injury (i.e., locked-in syndrome [LIS] and disorders of consciousness [DOC]). It is important to stress that a BCI is intended to be an alternative communication tool rather than a mind reader. A BCI provides subjects with a virtual keyboard, where keys are pressed by the modulation of brain activity (Sellers et al., 2010).

Signal acquisition – Human brain activity can be recorded either invasively, using intracortical microelectrode arrays (i.e., electrodes inserted inside the cortex [Hochberg et al., 2006, 2012]) or electrocorticography (i.e., electrodes placed directly on the surface of the cortex [Eliseyev and Aksenova, 2014]); or non-invasively, using EEG, fMRI, or functional near-infrared spectroscopy (i.e., local blood flow/oxygenation-related changes in absorption of near-infrared light [Naseer and Hong, 2015]). A user performs a specific task to modulate brain activity so that recordings captured simultaneously contain task-related information. Common tasks include imagining limb movements, performing complex arithmetic problems, and visualizing objects.

Feature extraction – Recorded brain activity is processed by a computer to remove task-unrelated activity and background noise. The remaining task-related activity is then used to compute features. The nature of these features depends on the imaging modality. For example, when using intracortical microelectrode arrays, the rate of action potentials per unit of time is commonly used as a feature. In fMRI, features may reflect activation in a given region of interest or functional connectivity between areas. In EEG, common features include characteristic voltage waveforms (i.e., evoked potentials) or computed power of EEG rhythms. Features can reflect brain responses evoked by stimuli, activation corresponding to motor volition (changes in activation of sensorimotor areas in response to a motor task), or modulation of brain activity learned by training (e.g., slow cortical potentials, defined as a progressive change in the mean EEG voltage across electrode[s]).

Classification – Before using the BCI as a communication tool, it needs to be trained to decode a user's intention. This may include user-training, in which the user learns to voluntarily modulate his/her brain activity to make a choice, and system-training (also known as a system calibration), in which the BCI learns to distinguish brain patterns linked to the user's intention. For the evaluation of response to command and communication in patients with DOC, it is important to develop a system requiring minimal or no user-training. Conversely, for neuro-feedback or rehabilitation applications, user training is of particular importance. In the former case, repeated activation of several brain areas across large windows of time are acquired and transmitted to an algorithm which looks for recurring patterns that are initially unknown. In the latter case, specific activation is desired (e.g., activation of a brain region of interest), and the user must learn to modulate it. The algorithms that interpret these changes in activation are continuous filters or discrete classifiers that learn patterns from past data, then recognize and label future brain activity. A classifier may use simple rules to determine the class of a trial. However, rules based on past brain activity cannot be easily determined in some BCI applications, particularly those involving brain-injured patients, where responses to stimuli and tasks are complex and not yet well-known. The classification accuracy computed on an independent test dataset is a frequently reported measure of the "efficacy" of a classifier. To translate this classifier efficacy to the context of DOC, it is important to keep in mind the different evaluations: response to commands and the communication. In the evaluation of response to commands, we aim to determine if the user is able to willfully respond to simple instruction (e.g., "imagine moving your right hand"), showing a goal-oriented conscious process, such that classifier efficiency represents the user's ability to produce specific changes in his/her brain activation. In the evaluation of communication, the system is first calibrated (this system-training step is similar to the classifier building in response to command evaluation) which is then used to decode the user response to simple yes/no questions with answer known a priori. In this case, we will report the number of correctly answered questions. In both type of evaluation, a permutation test is recommended to evaluate whether or not this accuracy is higher than an accuracy obtained by random fluctuations in the data, defined as the "chance level" (Noirhomme et al., 2014). In the context of conscious brain-injured patients, we want to evaluate the BCI's potential for use in daily communication. A common rule-of-thumb in the BCI community is that an accuracy of at least 70 percent is required in order for a communication tool to be useful (Kübler et al., 2009; Kübler and Birbaumer, 2008). However, from the perspective of patients with amyotrophic lateral sclerosis, a need for classification accuracy above 90 percent has been reported (Huggins et al., 2011).

Feedback and command – The BCI community commonly distinguishes between open-loop BCIs, in which neural data is recorded without any decoding while the BCI is in use, and closed-loop BCIs, in which neural data is decoded while the BCI is in use. In closed-loop, the decoder decision is used to provide feedback (e.g., visual, auditory, or tactile) to the user, allowing him/her to react to decoded results (in the event that intent is decoded incorrectly). Thus, in closed-loop BCIs, the decoded choice depends on the current neural activity and on previous neural history (Jarosiewicz et al., 2013). In this way, the BCI can provide access to a wide range of basic communication functions and environmental controls including email, text-to-speech engines, web browsers, home automation systems, etc. Through the computer, a user can operate any Windows-based program that can be operated with a keyboard (Sellers et al., 2010). Users can also control complex devices like neuroprostheses and mobile robots (or wheelchairs) (Millan et al., 2010).

A user's brain activity in the area of interest is recorded using fMRI, EEG, functional near-infrared spectroscopy, magnetoencephalography, electrocorticography, or an intracortical array. Data are pre-processed (e.g., spatial and/or temporal filtering) before task-related features can be extracted from the neural signal. A classifier is then built based on these features to recognize changes associated with the user's intention (i.e., defining different classes). Once trained, the classifier is used to select a specific class based on incoming neural data. The selected class is then fed back to the user allowing him/her to correct his/her choice. The result of the classification can also be used as a command to control an external machine (e.g., robotic arm, wheelchair)

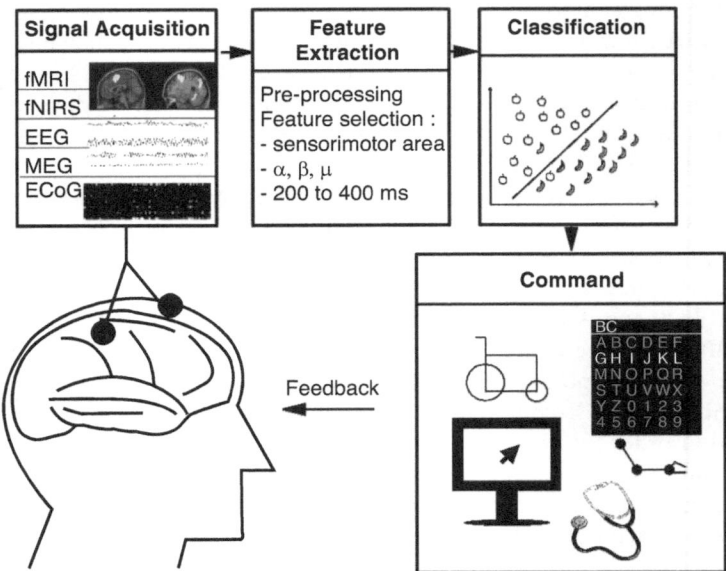

Figure 6.1 Illustration of a brain-computer interface loop scheme.
A color version of this image is available here: www.routledge.com/
products/9781138851672.

BCIs can be categorized as synchronous or asynchronous based on how
the system is being controlled by the user. A synchronous BCI is time-locked
to a cue and therefore necessitates that the user follows a specific scheme to
switch from one mental task to the next. On the other hand, an asynchronous
BCI is a self-paced system that can be controlled by the user independently
of a cue. Although asynchronous BCIs are more user-friendly, most of the
current BCIs are synchronous. The pace of a synchronous BCI is defined by
the system and can be either fixed or based on the system's ability to detect a
specific change in brain activity. For example, if a patient is asked to move his/
her hand, the signal can either be analyzed by looking at the whole trial (e.g.,
10"), or by using a moving window in order to detect a change in brain activity at any time during the trial (e.g., starting at 1", until 10"). The latter is
more comparable to what is done during bedside assessment of consciousness
(e.g., coma recovery scale-revised [CRS-R] [Giacino et al., 2004]). Indeed, a
certain amount of time is provided to the patient to follow a command, and as
soon as a response is observed within this time period, the behavior is scored
as present.

Despite their name, some BCIs are not fully motor-independent. For example, a majority of the visual paradigms used in BCI studies depend on gaze

control (i.e., the user has to move their eyes to focus on the correct stimulus in order to select it). For patients with eye control impairment (e.g., some cases of LIS [Bauer et al., 1979]), a BCI independent of any motor control is necessary.

A promising communication tool for the motor-disabled

The successful use of BCIs by a large population of healthy users has been reported in several studies (Guger et al., 2003, 2009, 2013). Furthermore, many studies have shown that these systems are feasible and practical for patient groups. Interestingly, a review study including 58 patients tested with different kinds of BCIs showed that there is no relationship between physical impairment and BCI performance (Kübler and Birbaumer, 2008). BCIs are used in rehabilitation for patients after a stroke, to restore movement in patients with para/tetraplegia by operating prosthetic devices, to allow patients with amyotrophic lateral sclerosis and LIS to communicate, and, as recently proposed, to help treat patient with psychiatric disorders through neurofeedback sessions. Most BCIs used in patients rely on EEG signals. EEG is a common neurophysiological modality available in most hospitals and research centers and enables testing of a range of paradigms in a wide population.

The first BCI study involving patients with severe motor impairment was reported by (Birbaumer et al., 1999) using slow cortical potentials to drive a spelling device. Slow cortical potentials with a negative polarity are usually observed during movement preparation. Patients watched a cursor move up and down depending on their brain activity, and they were instructed to modulate this activity in order to improve their control over the cursor. They were trained for hundreds of sessions over weeks before being able to communicate with the system.

The slow cortical potential BCI established interest in, and applicability of, BCI systems in pathological populations, but a more recent meta-analysis of 35 patients (25 of whom used slow cortical potential BCIs) reported that a switch to other BCI systems (based on P3 or motor imagery) was desirable in order to reduce the time needed for training (Kübler and Birbaumer, 2008). A P3 is a positive deflection in the EEG typically observed when users attend to a random sequence of rare target stimuli interspersed with frequent non-target stimuli. Most P3-BCIs use a 6x6 matrix in which each of the 36 cells contains a letter or symbol. The cells are then highlighted individually, by row/column, or following a predefined pattern. The user is asked to concentrate on a symbol that will be highlighted up to 30 times in half a minute depending on patient capabilities. Since the P3 is an automatic brain response, no user-training is required. Moreover, a small number of trials (five; see Guger et al., 2009) is sufficient to calibrate the system (i.e., to enable the system to define the electrodes and time window most appropriate for the user). Furthermore, P3-BCIs can easily be adapted to present a larger number of symbols. In addition to communication, P3-BCIs have been used for browsing the Internet and painting (Münßinger et al., 2010). The P3-speller is the most studied BCI so far (Cecotti, 2011). Mean accuracy in healthy individuals is about 90 percent, and 90 percent of the healthy users can use it with more than 80 percent accuracy (Guger et al., 2009). However, a recent review on the use of the P3-speller BCI showed a drop in accuracy to about 74 percent in patients (Marchetti and Priftis, 2014). One patient with amyotrophic lateral sclerosis successfully used a visual P3-BCI to manage his research lab for more than two and a half years (at the time of publication) (Sellers et al., 2010). However, one of the drawbacks of the visual P3-BCI is that it relies on the ability of the users to move their eyes to focus on the target symbol (Brunner et al., 2010; Treder and Blankertz, 2010). Additionally, most of the patients who are able to control a visual P3-BCI could use an eye-tracking system faster and with higher accuracy. For patients with limited eye control, a paradigm using single symbol presentation has been developed (Hoffmann et al., 2008). Another alternative is the use of a different modality, such as audition. In the first auditory P3-BCI proposed in 2006, the user was asked to concentrate on one of four different stimuli: yes, no, pass, end. An average accuracy of 71 percent was reported in three patients with amyotrophic lateral sclerosis (Sellers and Donchin, 2006). However, another study using a 25-choice auditory BCI, reported that only one out of three LIS patients could reach an accuracy above chance-level (Kübler et al., 2009). When compared to visual P3-BCIs in healthy individuals, for the same number of stimuli, auditory-P3 BCIs show a drop in accuracy (Furdea et al., 2009). Finally, a P3 can also be elicited using somatosensory stimulation. In the cohort included in this study, one out of six LIS patients could communicate using a two-choice vibrotactile stimulation BCI (Lugo et al., 2014). Further studies are needed to determine whether or not this approach can be improved upon.

Some researchers have focused on another type of event-related potential, the steady-state visually evoked potential (SSVEP). This neural oscillation, which synchronize with the stimulus frequency (and its harmonics), can be recorded using electrodes placed on the occipital and parieto-occipital areas. The advantage of SSVEP is that it is less susceptible to ocular and electromyographic artifacts. Additionally, a study including seven patients with LIS reported an accuracy above 70 percent achieved by all patients using an SSVEP-BCI, while only three were able to reach that level using a visual P3-BCI. Moreover, a lower mental workload and a higher overall satisfaction were reported with the SSVEP-BCI (Combaz et al., 2013). The main drawback with the SSVEP approach remains the dependency on gaze control. A proposed gaze-independent SSVEP-BCI demonstrated good performance in healthy volunteers, but lower performance in LIS patients. Specifically, only one out of four patients were able to communicate (performance above 70 percent) (Lesenfants et al., 2014).

P3 and SSVEP-based BCIs are stimulus-driven and rely on the user's perception of those stimuli. On the contrary, BCIs based on motor imagery use brain signals that are intentionally produced by the user. Changes in somatosensory motor rhythms, associated with the imagination of movements or their attempted execution, can be recorded over the primary somatosensory and motor cortical areas. Often, the number of imagined movements is limited to two. Based on several studies, eight out of the 14 patients tested could gain significant BCI control (70 to 80 percent accuracy) using motor imagery (Neuper et al., 2003; Kübler et al., 2005; Kübler and Birbaumer, 2008; Hohne et al., 2014). For one severely impaired patient (tetraparesis after cerebral bleeding), the motor imagery BCI outperformed his existing communication system (button press with residual movement of left thumb) in terms of control accuracy, reaction time and information transfer rate (Hohne et al., 2014). In addition, motor imagery BCIs offer further possibilities in the context of neurorehabilitation of patients. They may improve motor learning and motor rehabilitation by detecting intended movements and providing feedback (Rupp, 2014). By practicing motor imagery of the paralyzed limb with BCI feedback, stroke patients and patients with spinal cord injuries may better preserve the integrity of cortical neuronal connections or enhance neurological recovery of motor function (Kaiser et al., 2014).

While non-invasive methods are often limited to discrete decoding and short-term use, invasive recordings, such as intracortical BCI, are envisioned for long-term use and fine continuous control in some specific patient populations (e.g., patients with amyotrophic lateral sclerosis or spinal cord injuries). The BrainGate team (Brown University, Massachusetts General Hospital, Stanford University, Case Western Reserve University, and Providence VA Medical Center) have shown that a person with tetraplegia resulting from cervical spinal cord injury is able to control external devices such as a computer cursor using neural signals, simply by imagining movements of his own hand (Hochberg et al., 2006). Signals were recorded with a 4x4 mm array of 100 silicon microelectrodes (Blackrock Microsystems, Salt Lake City, Utah) neurosurgically implanted in the hand-arm area of the dominant primary motor cortex. Another patient with incomplete LIS was also able to control a robotic arm and serve herself a drink of coffee for the first time in nearly 15 years (Hochberg et al., 2012).

In addition to assisting patients lacking motor control, a growing field of research in BCI relates to biofeedback of neuronal responses, commonly called neurofeedback. In neurofeedback experiments, users learn to regulate their brain activity via accurate and rapid feedback of the measured brain activity (Weiskopf, 2012). The feedback is based on indirect measures of activation in the targeted area while using fMRI or functional near-infrared spectroscopy, or on the power of given frequency bands while using EEG. Neurofeedback has been proposed as an intervention in treating symptoms of various pathological states including Parkinson's disease, chronic pain, schizophrenia, Attention Deficit Hyperactivity Disorder, depression, and substance abuse.

Application of brain-computer interfaces in disorders of consciousness

fMRI-based BCI in the DOC

Motor imagery – BCI technology appeared for the first time in the context of DOC in 2006 when Owen et al. proposed a motor imagery paradigm, composed of two different tasks ("imagine you are playing tennis" and "imagine yourself walking through your house"), to a patient clinically diagnosed to be in a vegetative state/unresponsive wakefulness syndrome (VS/UWS) (Owen et al., 2006). Despite this patient being clinically diagnosed as unconscious, she showed task-related cognitive responses during consecutive trials similar to those observed in a cohort of healthy volunteers. Later on, this patient evolved to a minimally conscious state (MCS). This protocol was then evaluated on a cohort of 54 patients (23 VS/UWS and 31 MCS) in a follow-up study by (Monti et al., 2010). Four patients, clinically diagnosed to be in VS/UWS, could willfully modulate their brain activity according to the task. One MCS patient could also respond to command with this protocol. In particular, one of these five patients could also answer yes/no questions by concentrating on one task to answer "yes," and the other one for "no." Communication with this paradigm has also been replicated in a recent study (Fernandez-Espejo and Owen, 2013). In this study, a patient diagnosed to be in VS/UWS for more than 12 years could answer basic autobiographical, semantic, orientation, new knowledge, and personal preference questions, as well as questions about quality of life. In 2011, Bardin et al. evaluated the ability of patients with DOC to modulate their brain activity using another motor imaginary task (imagining swimming or playing tennis) (Bardin et al., 2011). Three out of the six patients included were able to follow the command successfully.

Auditory modalities – In 2009, Monti, Coleman, and Owen proposed assessing residual executive functions with an fMRI paradigm aimed at evaluating the preserved cognitive function without requiring patients to produce any behavioral output (Monti et al., 2009). One MCS with bedside response to command was able to count a target word in an auditory sequence of non-targets word and showed preserved working memory abilities exceeding that which could be observed with standard behavioral assessment. In another study, three patients (one VS/UWS and two MCS) were instructed to either count the occurrences of a target word ("yes" or "no") or to simply relax and passively listening to a sequence of "yes" and "no" interspersed within a random series of numbers (Naci and Owen, 2013). All three patients were able to auditory follow the command, and two of them were able to answer simple yes/no question (e.g., "are you in a supermarket?") with this protocol.

EEG-based BCI in the DOC

Despite the many advantages of fMRI, this technique is limited in terms of availability, affordability, and ease of use in this population. Moreover, patients with metallic implants cannot be evaluated with an fMRI paradigm: non-ferrous implants will produce image artifacts while ferrous metallic implants are a contraindication to MRI. Patient's head motions (reflexive movements in the scanner, general restlessness, or decreased patient cooperation) are also an important limitation of MRI, resulting in uninterpretable data. These exclusion criteria have been observed during the European FP7 DECODER project (deployment of brain-computer interfaces for the Detection of Consciousness in Non-Responsive Patients) during which only 60 out of the 169 DOC patients selected for an fMRI active sport/navigation paradigm at the Centre Hospitalier Universitaire de Liège (Belgium) could produce interpretable data. As a results, in recent years, researchers have started to focus on the development of EEG-based active tasks to assess response to command in DOC. EEG is accessible in most clinical settings, is compact and inexpensive, and can be readily deployed at the patients' bedsides without contraindications to metallic implants, or high sensitivity to motion.

Motor imagery – Following fMRI studies based on motor imagery, (Goldfine et al., 2011) developed an EEG paradigm using motor imagery (swimming) and a spatial navigation task. Significant differences in EEG activity between the two conditions could be observed in two out of the three patients showing the ability to follow commands at the bedside (one MCS and one LIS patients). Distinctions between foot and hand movements ("imagine squeezing your right hand" versus "imagine moving all your toes") have also been observed in VS/UWS (Cruse et al., 2011) and in MCS patients (Cruse et al., 2012a) using a block paradigm (trials presented in block of 15 of the same type). Eight patients (3/16 VS/UWS, 5/23 MCS) were able to voluntary control their brain activity in response to commands. Recently, Coyle et al. (2015) reported patients with willful modulation of sensorimotor rhythms in real-time (imagined hand movement or toe wiggling) using the same block paradigm. Visual and/or auditory feedback was used to encourage alertness and motivation. Consistent activation was observed in multiple sessions for the four MCS patients included in the study. Using a trial based paradigm, one patient with VS/UWS was able to modulate brain activity with both EEG and fMRI in response to commands (Cruse et al., 2012b; Gibson et al., 2014b). Two other behaviorally unresponsive patients responded to commands with the fMRI paradigm only (Gibson et al., 2014b).

Auditory modalities – Responses to auditory stimuli have also been investigated in active EEG paradigms. Schnakers et al. (2008) proposed an auditory P3-BCI and asked the 22 patients included to count the number of times a name (patients' own name or unfamiliar names) was presented within an auditory sequence of random names. Nine out of the 14 MCS patients included in this study showed larger P3 responses during active own (five patients) or unfamiliar name counting than passive listening. None of the VS/UWS patients showed a response to the active counting task. It is important to stress that a patient, behaviorally diagnosed to be in coma, showed a significant difference in response between the passive and the active tasks, and was subsequently reassessed and diagnosed with a complete LIS (Schnakers et al., 2009). More recently, Risetti et al. evaluated Schnakers' auditory P3 paradigm on 11 patients with DOC (8 VS/UWS and 3 MCS) and showed similar results (Risetti et al., 2013). Chennu et al. (2013) used a task designed to engender exogenous or endogenous attention – indexed by the P3a and P3b components, respectively – in response to a pair of auditory word stimuli presented amongst distracters. These two subcomponents of the P3 response to auditory stimuli have been shown to be elicited by "bottom-up" stimulus novelty that may be task-irrelevant (P3a) and "top-down" volitional engagement (P3b) of endogenous attention to task-relevant targets (Polich, 2007). Out of the 21 patients, one VS/UWS patient illustrated P3a and P3b responses while two MCS patients showed only a P3a. Interestingly, 20 of these patients were also administered the fMRI paradigm developed by Owen et al. (Owen et al., 2006; Monti et al., 2010). In six patients in whom no discernible P3a/P3b response could be elicited, a response to command using the fMRI tennis imagery task could be detected. The VS/UWS patient who showed P3a/P3b responses did also show a response to command with the fMRI, suggesting that the presence of a P3a and P3b may highlight a well-preserved volitional attention process. A study by Lule et al. (2013) used a four-choice (yes, no, stop, go) auditory based-paradigm with three VS/UWS, 13 MCS, and two LIS patients. In this auditory paradigm, the evaluation of offline response to command ("concentrate on repetitions of 'yes' or 'no' presented in the stream of the four different words") and online yes/no communication ("concentrate on the repetition of 'yes' to answer 'yes' to the question") were performed. Only one MCS and one LIS patients showed offline performance above chance-level. None of the patients illustrated functional communication. Finally, Pokorny et al. (2013) tested an auditory P3-based BCI based on tone stream segregation allowing for binary decisions in 12 MCS patients. Two tone streams with infrequently and randomly appearing deviant tones were presented to the patient. The patient was asked to count the number of deviants in one stream in order to elicit a P3 response in the attended stream. Out of the 12 patients included, five could achieve performance above chance-level on a single-trial basis (nine out of the 12 patients after averaging across all data and computing significant differences).

Hybrid modalities – Combinations of multiple brain responses have been suggested by Pan et al. (2014) in the context of disorders of consciousness. In this study, the patient's own face and an unfamiliar face were randomly displayed on the left and right side of a computer screen. The left and right images were flickering from appearance to disappearance at different frequencies while the two images also flashed from appearance to disappearance in a random order, eliciting both SSVEP and P3 responses. One of the four VS/UWS, one of the three MCS, and the LIS patient were able to selectively attend to their own or the unfamiliar image. Two other patients (one VS/UWS and one MCS) failed to attend to the unfamiliar image but achieved accuracies above chance-level for their own image. Interestingly, none of the patients with DOC showed a response to command behaviorally at the bedside.

Achilles' heel and closer look toward the future

A good knowledge of the system proposed and the population studied is essential when developing a new BCI-based diagnostic tool. Indeed, the successful translation of BCI technology from healthy individuals to severely brain-injured patients presents a considerable number of challenges and limitations that should be overcome in future studies to improve current performance.

First, the patient's general condition is a constraint to take into account when developing a new paradigm:

1 **Vigilance fluctuation** – Patients with altered states of consciousness may present fluctuations in their level of vigilance over time. It is crucial to distinguish a conscious patient who is unable to follow commands due to fluctuations in vigilance from an unconscious patient. BCI evaluations must be repeated across days to ensure a reliable diagnosis and take into account these inter-session fluctuations, as recommended for CRS-R evaluation. This would require portable systems available at the bedside. During a session, identifying and rejecting trials with low-levels of vigilance in real-time, prior to classification, would allow for results that are independent of vigilance fluctuations.

2 **Motivation** – The success of a BCI paradigm also relies upon the patient's willingness or motivation to perform a task (Kleih et al., 2010; Nijboer et al., 2010) or akinetic mutism (Giacino, 1997). Those factors must be considered with care as we cannot distinguish a patient lacking motivation to do the task and an unconscious patient. Providing feedback through a BCI system can help increase motivation (Coyle et al., 2015; Lule et al., 2013), as can improving patient engagement through motivating/goal-oriented tasks (e.g., a game or a bell to call a nurse). Patient-tailored tasks should also be considered since tailoring paradigms to a patient's previous habits has been shown to help increase BCI sensitivity (Gibson et al., 2014a). However, this would require flexible paradigms that can be modified based on new patients' preferences.

3 **Fatigue** – DOC patients are easily exhausted and can present limited attention spans and memory capacities. This means that paradigms that succeed with healthy users may not necessarily succeed with these patients. For patients, sessions should be short and task complexity/cognitive workload should be low. Communication evaluation should be assessed with simple questions with answers known a priori since severely brain-damaged patients may have difficulty giving accurate answers otherwise (Nakase-Richardson et al., 2009). Block designs have been proposed to reduce the cognitive load required to complete tasks (Cruse et al., 2011; Coyle et al., 2015). However, in DOC patient populations, block designs should be avoided because changes in the EEG signal across and within blocks may be influenced by vigilance and motor artifacts leading to a lack of independence between trials (non-stationarities; see Goldfine et al., 2013; Henriques et al., 2014). In the study of Chennu et al. (2013), the difference in task complexity could probably explain the

difference in the performance achieved by P3 EEG-BCI as compared to motor imagery fMRI-BCI.

4   **Sensory deficit** – Following brain injury, DOC patients classically suffer from severe sensory deficit. Motor control is often impaired or non-existent, preventing the use of motor-dependent BCI paradigms. As such, gaze-independent visual paradigms must be considered (Lesenfants et al., 2014). In addition to this motor control deficits, several studies have suggested that motor imagery cannot be reliably used and interpreted in severely motor-disabled patients (Bai et al., 2008; Kübler et al., 2005). Instead of motor imagery, Nijboer et al. (2010) recommended the use of the P3 response in patients with severe motor impairment. The key challenge is therefore to develop reliable systems offering stimuli, instruction, and/or question presentation through multiple modalities. Hybrid BCIs could also help improve performance by combining multiple brain responses or modalities (Pan et al., 2014). Future systems should also incorporate a passive evaluation of sensory pathways prior to the active task to help select the most appropriate modalities.

5   **Neuroplasticity** – Following a brain injury, cortical pathways and synapses are likely to remap. This cortical remapping increases difficulty when translating a protocol from healthy individuals to brain-injured patients (Chennu et al., 2013; Nam et al., 2012). Distribution of electrodes around the area of interest, but also surrounding this area, could avoid missing brain responses developed in newly allocated area. Then, to optimize the performance, the reduction of the number of electrodes based on relevance to decoding should be considered. An example of this translation could be observed in Pokorny et al. (2013) in which the auditory paradigm, first developed in healthy controls, had to be adapted for use by patients with DOC.

6   **Aphasia** – Language impairments, such as aphasia, have high prevalence in brain-injured population, ranging from 15 to 30 percent (Inatomi et al., 2008; Laska et al., 2001; Eisenberg et al., 1990; Chapman et al., 1995). However, EEG-BCIs proposed in the literature are often based on instructions that have to be understood to successfully achieve the task, meaning that they are potentially inappropriate for conscious patients suffering from language impairments (Majerus et al., 2009). In such cases, language-independent paradigms will be needed (for a review, see Boly and Seth, 2012).

Second, the BCI should be adapted to this population based on its application in a patient's day-to-day life:

1   **Signal quality** – Patients' motion is difficult to predict or control, leading to suboptimal data quality. These confounding factors have to be overcome with the assistance of appropriate statistical analyses to reduce false negatives and positives (Cruse et al., 2013; Goldfine et al., 2013). Moreover, respiratory and nutrition systems are often present in the patients' rooms, increasing electrical noise. Selection of high signal-to-noise modalities such as the Steady-State Visually Evoked Potentials (SSVEP), along with artifact-rejection

algorithms, could help robustness in challenging environments far away from the laboratory.

2 **Calibration** – Classifying response to command requires the user to modulate his/her brain activity on a majority of task-trials to ensure performance above chance level. However, several parameters, such as fluctuation of vigilance and fatigue, could limit the percentage of successfully completed trials. Ideally, single-trial evaluation would provide sufficiently high confidence in user response, but this is currently not the case. Trial rejection methods, such as Spatial-Temporal Discriminant Analysis (Zhang et al., 2013) and Step-Wise Linear Discriminant Analysis (Krusienski et al., 2006), have been proposed to reduce the effect of limited attention and maximize the discriminant information between classes.

3 **Long-term use** – EEG-BCIs are adapted to evaluate communication and response to command during intermittent sessions in clinical routine. However, for patients diagnosed as conscious, long-term EEG-based BCI communication is limited by the need for conductive gel. Indeed, daily replacement of EEG gel would potentially limit user-independence. Dry electrodes have been proposed as an alternative to this inconvenience, but they suffer from high artifact sensitivity. Brain-computer interfaces using electrocorticography or intracortical signals could overcome this limitation, and illustrate good spatial and temporal resolutions as well as high signal-to-noise ratio. Recent research in these fields has demonstrated the ability of these technologies to restore point-and-click computer-cursor and robotic arm control in people with tetraplegia (Hochberg et al., 2006, 2012).

4 **Results** – Currently, there is still no consensus in the statistical methods used to define significant results. In fact, some results have been controverted following reanalysis of the data (Goldfine et al., 2013; Noirhomme et al., 2014). Margins must be defined to avoid false-positive close-to-chance accuracies and increase reliability of the system. Moreover, BCI studies should report standardized measures of diagnostic tool quality such as false-positive and false-negative ratios. Unfortunately, despite promising results in the BCI literature in the context of DOC, the observed false negative rates remain high with every technique, ranging from 25 to 100 percent (see Figure 6.2). The low-sensitivity to conscious patients showing behavioral signs of consciousness is often pushed into the background, giving greater importance to miracle cases. However, a system that is not sensitive to detecting patients diagnosed as conscious with behavioral scales at the bedside could neither be reliably used in patients with unclear diagnoses nor be trusted in positive detection of unconscious patients by clinicians (Pazart et al., 2015).

Only studies with at least five patients responding to command at the bedside and patients' clinical information were included. The number of patients included in each study is provided on top of each bar. Colors refer to the paradigm used. We decomposed the false-negative ratio of each modality

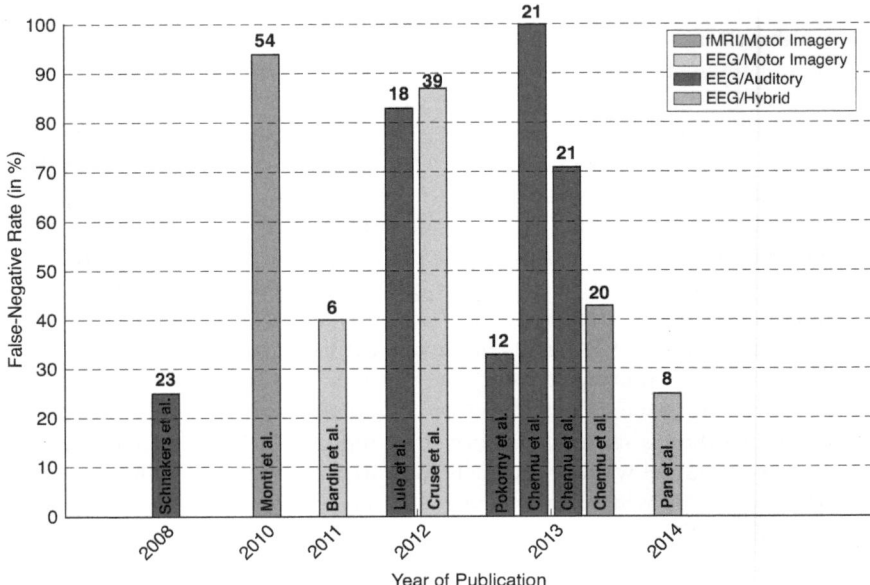

Figure 6.2 Studies using brain-computer interfaces in the context of disorders of consciousness with their respective false-negative ratios (i.e., the percentage of patients responding behaviorally to commands that cannot be detected by the BCI).

evaluated in Chennu et al. (2013). The false-negative ratio of Pokorny et al. (2013) is based on CRS-R values obtained from the authors. Note that for four patients, subscale scores were missing, limiting the current analysis in terms of false negatives.

Overcoming these challenges could lead to reliable BCI-based diagnostic tools adapted to the DOC context. Nevertheless, guidelines for clinicians, in terms of interpretation and use of findings, must define and account for BCI limitations (e.g., false-negative rate, false-positive rate, chance-level). These guidelines will influence discussions with families, patient care (i.e., pain management), end-of-life decisions, and potential user-constraints (e.g., cortical deafness, oculomotor impairments).

Conclusion

In this chapter, we reviewed the current BCI literature in the context of brain-injured patients. Improving diagnoses and unlocking communication ability in these patients could lead to improved rehabilitation strategies, quality of life, and prognosis. However, despite the increasing interest of the BCI community in working with DOC patients, high false-negative and false-positive rates illustrate the difficulty in translating this technology from healthy individuals to severe brain-injured patients. Disregarding the clinical reality of these patients, which can vary substantially from one person to another, would lead to unsuccessful BCI designs. Given the importance of an objective and motor-independent diagnostic tool for non-communicative brain-injured patients, the research community should push forward with its efforts to illuminate our knowledge of consciousness and unlock the locked.

Acknowledgment

This study was supported by the National Funds for Scientific Research (FNRS), European ICT Programme Projects FP7-247919 DECODER, FP7-602450 IMAGEMEND, French Speaking Community Concerted Research Action, University of Liege, the Belgian American Educational Foundation (BAEF), the Fédération Wallonie Bruxelles International (WBI), the Massachusetts General Hospital Department of Neurology and Division of Neurocritical Care and Emergency Neurology, and the James McDonnell Foundation.

References

Bai, O., Lin, P., Vorbach, S., Floeter, M. K., Hattori, N., & Hallett, M. 2008. A high performance sensorimotor beta rhythm-based brain-computer interface associated with human natural motor behavior. *J Neural Eng*, 5, 24–35.

Bardin, J. C., Fins, J. J., Katz, D. I., Hersh, J., Heier, L. A., Tabelow, K., Dyke, J. P., Ballon, D. J., Schiff, N. D., & Voss, H. U. 2011. Dissociations between behavioural and functional magnetic resonance imaging-based evaluations of cognitive function after brain injury. *Brain*, 134, 769–82.

Bauer, G., Gerstenbrand, F., & Rumpl, E. 1979. Varieties of the locked-in syndrome. *J Neurol*, 221, 77–91.

Birbaumer, N., Ghanayim, N., Hinterberger, T., Iversen, I., Kotchoubey, B., Kübler, A., Perelmouter, J., Taub, E., & Flor, H. 1999. A spelling device for the paralysed. *Nature*, 398, 297–8.

Boly, M., & Seth, A. K. 2012. Modes and models in disorders of consciousness science. *Arch Ital Biol*, 150, 172–84.

Brunner, P., Joshi, S., Briskin, S., Wolpaw, J. R., Bischof, H., & Schalk, G. 2010. Does the "P300" speller depend on eye gaze? *J Neural Eng*, 7, 056013.

Cecotti, H. 2011. Spelling with non-invasive Brain-Computer Interfaces – current and future trends. *J Physiol Paris*, 105, 106–14.

Chapman, S. B., Levin, H. S., & Culhane, K. A. 1995. Language impairment in closed head injury. In H. S. Kirshner (Ed.), *Handbook of neurological speech and language disorders*. New York: Informa Health Care.

Chennu, S., Finoia, P., Kamau, E., Monti, M. M., Allanson, J., Pickard, J. D., Owen, A. M., & Bekinschtein, T. A. 2013. Dissociable endogenous and exogenous attention in disorders of consciousness. *Neuroimage Clin*, 3, 450–61.

Combaz, A., Chatelle, C., Robben, A., Vanhoof, G., Goeleven, A., Thijs, V., Van Hulle, M. M., & Laureys, S. 2013. A comparison of two spelling Brain-Computer Interfaces based on visual P3 and SSVEP in Locked-In Syndrome. *PLoS One*, 8, e73691.

Coyle, D., Stow, J., Mccreadie, K., Mcelligott, J., & Carroll, A. 2015. Sensorimotor modulation assessment and brain-computer interface training in disorders of consciousness. *Arch Phys Med Rehabil*, 96, S62–70.

Cruse, D., Chennu, S., Chatelle, C., Bekinschtein, T. A., Fernandez-Espejo, D., Pickard, J. D., Laureys, S., & Owen, A. M. 2011. Bedside detection of awareness in the vegetative state: A cohort study. *Lancet*, 378, 2088–94.

Cruse, D., Chennu, S., Chatelle, C., Bekinschtein, T. A., Fernandez-Espejo, D., Pickard, J. D., Laureys, S., & Owen, A. M. 2013. Reanalysis of "Bedside detection of awareness in the vegetative state: A cohort study" – Authors' reply. *Lancet*, 381, 291–2.

Cruse, D., Chennu, S., Chatelle, C., Fernandez-Espejo, D., Bekinschtein, T. A., Pickard, J. D., Laureys, S., & Owen, A. M. 2012a. Relationship between etiology and covert cognition in the minimally conscious state. *Neurology*, 78, 816–22.

Cruse, D., Chennu, S., Fernandez-Espejo, D., Payne, W. L., Young, G. B., & Owen, A. M. 2012b. Detecting awareness in the vegetative state: Electroencephalographic evidence for attempted movements to command. *PLoS One*, 7, e49933.

Eisenberg, H. M., Gary, H. E., JR., Aldrich, E. F., Saydjari, C., Turner, B., Foulkes, M. A., Jane, J. A., Marmarou, A., Marshall, L. F., & Young, H. F. 1990. Initial CT findings in 753 patients with severe head injury. A report from the NIH Traumatic Coma Data Bank. *J Neurosurg*, 73, 688–98.

Eliseyev, A., & Aksenova, T. 2014. Stable and artifact-resistant decoding of 3D hand trajectories from ECoG signals using the generalized additive model. *J Neural Eng*, 11, 066005.

Fernandez-Espejo, D., & Owen, A. M. 2013. Detecting awareness after severe brain injury. *Nat Rev Neurosci*, 14, 801–9.

Furdea, A., Halder, S., Krusienski, D. J., Bross, D., Nijboer, F., Birbaumer, N., & Kübler, A. 2009. An auditory oddball (P300) spelling system for brain-computer interfaces. *Psychophysiology*, 46, 617–25.

Giacino, J. T. 1997. Disorders of consciousness: Differential diagnosis and neuropathologic features. *Semin Neurol*, 17, 105–11.

Giacino, J. T., Kalmar, K., & Whyte, J. 2004. The JFK Coma Recovery Scale-Revised: Measurement characteristics and diagnostic utility. *Arch Phys Med Rehabil*, 85, 2020–9.

Gibson, R. M., Chennu, S., Owen, A. M., & Cruse, D. 2014a. Complexity and familiarity enhance single-trial detectability of imagined movements with electro-encephalography. *Clin Neurophysiol*, 125, 1556–67.

Gibson, R. M., Fernandez-Espejo, D., Gonzalez-Lara, L. E., Kwan, B. Y., Lee, D. H., Owen, A. M., & Cruse, D. 2014b. Multiple tasks and neuroimaging modalities increase the likelihood of detecting covert awareness in patients with disorders of consciousness. *Front Hum Neurosci*, 8, 950.

Goldfine, A. M., Bardin, J. C., Noirhomme, Q., Fins, J. J., Schiff, N. D., & Victor, J. D. 2013. Reanalysis of "Bedside detection of awareness in the vegetative state: A cohort study." *Lancet*, 381, 289–91.

Goldfine, A. M., Victor, J. D., Conte, M. M., Bardin, J. C., & Schiff, N. D. 2011. Determination of awareness in patients with severe brain injury using EEG power spectral analysis. *Clin Neurophysiol*, 122, 2157–68.

Guger, C., Allison, B. Z., & Edlinger, G. 2013. *Brain-Computer Interface Research. A State-of-the-Art Summary*. Berlin: Springer-Verlag.

Guger, C., Daban, S., Sellers, E., Holzner, C., Krausz, G., Carabalona, R., Gramatica, F., & Edlinger, G. 2009. How many people are able to control a P300-based brain-computer interface (BCI)? *Neurosci Lett*, 462, 94–8.

Guger, C., Edlinger, G., Harkam, W., Niedermayer, I., & Pfurtscheller, G. 2003. How many people are able to operate an EEG-based brain-computer interface (BCI)? *IEEE Trans Neural Syst Rehabil Eng*, 11, 145–7.

Henriques, J., Gabriel, D., Grigoryeva, L., Haffen, E., Moulin, T., Aubry, R., Pazart, L., & Ortega, J. P. 2014. Protocol design challenges in the detection of awareness in aware subjects using EEG signals. *Clin EEG Neurosci*.

Hochberg, L. R., Bacher, D., Jarosiewicz, B., Masse, N. Y., Simeral, J. D., Vogel, J., Haddadin, S., Liu, J., Cash, S. S., Van Der Smagt, P., & Donoghue, J. P. 2012. Reach and grasp by people with tetraplegia using a neurally controlled robotic arm. *Nature*, 485, 372–5.

Hochberg, L. R., Serruya, M. D., Friehs, G. M., Mukand, J. A., Saleh, M., Caplan, A. H., Branner, A., Chen, D., Penn, R. D., & Donoghue, J. P. 2006. Neuronal ensemble control of prosthetic devices by a human with tetraplegia. *Nature*, 442, 164–71.

Hoffmann, U., Vesin, J. M., Ebrahimi, T., & Diserens, K. 2008. An efficient P300-based brain-computer interface for disabled subjects. *J Neurosci Methods*, 167, 115–25.

Hohne, J., Holz, E., Staiger-Salzer, P., Muller, K. R., Kübler, A., & Tangermann, M. 2014. Motor imagery for severely motor-impaired patients: Evidence for brain-computer interfacing as superior control solution. *PLoS One*, 9, e104854.

Huggins, J. E., Wren, P. A., & Gruis, K. L. 2011. What would brain-computer interface users want? Opinions and priorities of potential users with amyotrophic lateral sclerosis. *Amyotroph Lateral Scler*, 12, 318–24.

Inatomi, Y., Yonehara, T., Omiya, S., Hashimoto, Y., Hirano, T., & Uchino, M. 2008. Aphasia during the acute phase in ischemic stroke. *Cerebrovasc Dis*, 25, 316–23.

Jarosiewicz, B., Masse, N. Y., Bacher, D., Cash, S. S., Eskandar, E., Friehs, G., Donoghue, J. P., & Hochberg, L. R. 2013. Advantages of closed-loop calibration in intracortical brain-computer interfaces for people with tetraplegia. *J Neural Eng*, 10, 046012.

Kaiser, V., Bauernfeind, G., Kreilinger, A., Kaufmann, T., Kübler, A., Neuper, C., & Muller-Putz, G. R. 2014. Cortical effects of user training in a motor imagery based brain-computer interface measured by fNIRS and EEG. *Neuroimage*, 85 Pt 1, 432–44.

Kleih, S. C., Nijboer, F., Halder, S., & Kübler, A. 2010. Motivation modulates the P300 amplitude during brain-computer interface use. *Clin Neurophysiol*, 121, 1023–31.

Krusienski, D. J., Sellers, E. W., Cabestaing, F., Bayoudh, S., Mcfarland, D. J., Vaughan, T. M., & Wolpaw, J. R. 2006. A comparison of classification techniques for the P300 Speller. *J Neural Eng*, 3, 299–305.

Kübler, A., & Birbaumer, N. 2008. Brain-computer interfaces and communication in paralysis: Extinction of goal directed thinking in completely paralysed patients? *Clin Neurophysiol*, 119, 2658–66.

Kübler, A., Furdea, A., Halder, S., Hammer, E. M., Nijboer, F., & Kotchoubey, B. 2009. A brain-computer interface controlled auditory event-related potential (p300) spelling system for locked-in patients. *Ann N Y Acad Sci*, 1157, 90–100.

Kübler, A., Nijboer, F., Mellinger, J., Vaughan, T. M., Pawelzik, H., Schalk, G., Mcfarland, D. J., Birbaumer, N., & Wolpaw, J. R. 2005. Patients with ALS can use sensorimotor rhythms to operate a brain-computer interface. *Neurology*, 64, 1775–7.

Laska, A. C., Hellblom, A., Murray, V., Kahan, T., & Von Arbin, M. 2001. Aphasia in acute stroke and relation to outcome. *J Intern Med*, 249, 413–22.

Lesenfants, D., Habbal, D., Lugo, Z., Lebeau, M., Horki, P., Amico, E., Pokorny, C., Gomez, F., Soddu, A., Muller-Putz, G., Laureys, S., & Noirhomme, Q. 2014. An independent SSVEP-based brain-computer interface in locked-in syndrome. *J Neural Eng*, 11, 035002.

Lugo, Z. R., Rodriguez, J., Lechner, A., Ortner, R., Gantner, I. S., Laureys, S., Noirhomme, Q., & Guger, C. 2014. A vibrotactile p300-based brain-computer interface for consciousness detection and communication. *Clin EEG Neurosci*, 45, 14–21.

Lule, D., Noirhomme, Q., Kleih, S. C., Chatelle, C., Halder, S., Demertzi, A., Bruno, M. A., Gosseries, O., Vanhaudenhuyse, A., Schnakers, C., Thonnard, M.,

Soddu, A., Kübler, A., & Laureys, S. 2013. Probing command following in patients with disorders of consciousness using a brain-computer interface. *Clin Neurophysiol*, 124, 101–6.

Majerus, S., Bruno, M. A., Schnakers, C., Giacino, J. T., & Laureys, S. 2009. The problem of aphasia in the assessment of consciousness in brain-damaged patients. *Prog Brain Res*, 177, 49–61.

Marchetti, M., & Priftis, K. 2014. Effectiveness of the P3-speller in brain-computer interfaces for amyotrophic lateral sclerosis patients: A systematic review and meta-analysis. *Front Neuroeng*, 7, 12.

Millan, J. D., Rupp, R., Muller-Putz, G. R., Murray-Smith, R., Giugliemma, C., Tangermann, M., Vidaurre, C., Cincotti, F., Kübler, A., Leeb, R., Neuper, C., Muller, K. R., & Mattia, D. 2010. Combining brain-computer interfaces and assistive technologies: State-of-the-art and challenges. *Front Neurosci*, 4.

Monti, M. M., Coleman, M. R., & Owen, A. M. 2009. Executive functions in the absence of behavior: Functional imaging of the minimally conscious state. *Prog Brain Res*, 177, 249–60.

Monti, M. M., Vanhaudenhuyse, A., Coleman, M. R., Boly, M., Pickard, J. D., Tshibanda, L., Owen, A. M., & Laureys, S. 2010. Willful modulation of brain activity in disorders of consciousness. *N Engl J Med*, 362, 579–89.

Münßinger, J. I., Halder, S., Kleih, S. C., Furdea, A., Raco, V., Hösle, A., & Kübler, A. 2010. Brain painting: First evaluation of a new brain-computer interface application with ALS-patients and healthy volunteers. *Front Neurosci*, 4, 182.

Naci, L., & Owen, A. M. 2013. Making every word count for nonresponsive patients. *JAMA Neurol*, 70, 1235–41.

Nakase-Richardson, R., Yablon, S. A., Sherer, M., Nick, T. G., & Evans, C. C. 2009. Emergence from minimally conscious state: Insights from evaluation of post-traumatic confusion. *Neurology*, 73, 1120–6.

Nam, C. S., Woo, J., & Bahn, S. 2012. Severe motor disability affects functional cortical integration in the context of brain-computer interface (BCI) use. *Ergonomics*, 55, 581–91.

Naseer, N., & Hong, K. S. 2015. fNIRS-based brain-computer interfaces: A review. *Front Hum Neurosci*, 9, 3.

Neuper, C., Muller, G. R., Kübler, A., Birbaumer, N., & Pfurtscheller, G. 2003. Clinical application of an EEG-based brain-computer interface: A case study in a patient with severe motor impairment. *Clin Neurophysiol*, 114, 399–409.

Nijboer, F., Birbaumer, N., & Kübler, A. 2010. The influence of psychological state and motivation on brain-computer interface performance in patients with amyotrophic lateral sclerosis – a longitudinal study. *Front Neurosci*, 4.

Noirhomme, Q., Lesenfants, D., Gomez, F., Soddu, A., Schrouff, J., Garraux, G., Luxen, A., Phillips, C., & Laureys, S. 2014. Biased binomial assessment of cross-validated estimation of classification accuracies illustrated in diagnosis predictions. *Neuroimage Clin*, 4, 687–94.

Owen, A. M., Coleman, M. R., Boly, M., Davis, M. H., Laureys, S., & Pickard, J. D. 2006. Detecting awareness in the vegetative state. *Science*, 313, 1402.

Pan, J., Xie, Q., He, Y., Wang, F., Di, H., Laureys, S., Yu, R., & Li, Y. 2014. Detecting awareness in patients with disorders of consciousness using a hybrid brain-computer interface. *J Neural Eng*, 11, 056007.

Pazart, L., Gabriel, D., Cretin, E., & Aubry, R. 2015. Neuroimaging for detecting covert awareness in patients with disorders of consciousness: Reinforce the place of clinical feeling! *Front Hum Neurosci*, 9, 78.

Pokorny, C., Klobassa, D. S., Pichler, G., Erlbeck, H., Real, R. G., Kübler, A., Lesenfants, D., Habbal, D., Noirhomme, Q., Risetti, M., Mattia, D., & Muller-Putz, G. R. 2013. The auditory P300-based single-switch brain-computer interface: Paradigm transition from healthy subjects to minimally conscious patients. *Artif Intell Med*, 59, 81–90.

Polich, J. 2007. Updating P300: An integrative theory of P3a and P3b. *Clin Neurophysiol*, 118, 2128–48.

Risetti, M., Formisano, R., Toppi, J., Quitadamo, L. R., Bianchi, L., Astolfi, L., Cincotti, F., & Mattia, D. 2013. On ERPs detection in disorders of consciousness rehabilitation. *Front Hum Neurosci*, 7, 775.

Rupp, R. 2014. Challenges in clinical applications of brain computer interfaces in individuals with spinal cord injury. *Front Neuroeng*, 7, 38.

Schnakers, C., Perrin, F., Schabus, M., Hustinx, R., Majerus, S., Moonen, G., Boly, M., Vanhaudenhuyse, A., Bruno, M. A., & Laureys, S. 2009. Detecting consciousness in a total locked-in syndrome: An active event-related paradigm. *Neurocase*, 15, 271–7.

Schnakers, C., Perrin, F., Schabus, M., Majerus, S., Ledoux, D., Damas, P., Boly, M., Vanhaudenhuyse, A., Bruno, M. A., Moonen, G., & Laureys, S. 2008. Voluntary brain processing in disorders of consciousness. *Neurology*, 71, 1614–20.

Sellers, E. W., & Donchin, E. 2006. A P300-based brain-computer interface: Initial tests by ALS patients. *Clin Neurophysiol*, 117, 538–48.

Sellers, E. W., Vaughan, T. M., & Wolpaw, J. R. 2010. A brain-computer interface for long-term independent home use. *Amyotroph Lateral Scler*, 11, 449–55.

Treder, M. S., & Blankertz, B. 2010. (C)overt attention and visual speller design in an ERP-based brain-computer interface. *Behav Brain Funct*, 6, 28.

Weiskopf, N. 2012. Real-time fMRI and its application to neurofeedback. *Neuroimage*, 62, 682–92.

Wolpaw, J. R., Birbaumer, N., Mcfarland, D. J., Pfurtscheller, G., & Vaughan, T. M. 2002. Brain-computer interfaces for communication and control. *Clin Neurophysiol*, 113, 767–91.

Zhang, Y., Zhou, G., Zhao, Q., Jin, J., Wang, X., & Cichocki, A. 2013. Spatial-temporal discriminant analysis for ERP-based brain-computer interface. *IEEE Trans Neural Syst Rehabil Eng*, 21, 233–43.

7 Does task-evoked activity entail consciousness in vegetative state?

"Neuronal-phenomenal inference" versus "neuronal-phenomenal dissociation"

Georg Northoff

Neuroempirical background: is consciousness based on cognition?

Various imaging studies have been conducted during passive sensory stimulation, using mostly auditory, somatosensory, and visual stimuli (see Laureys and Schiff, 2012, for an overview). Most of these studies show somehow preserved activation in auditory and visual cortex in VS, though on a lower level compared to that in minimally conscious state (MCS) and healthy subjects. More specifically, MCS patients show a more widespread activation and higher degrees of long-range functional connectivity in midline regions and lateral fronto-parietal cortex than in VS patients.

These earlier sensory-based studies have recently been complemented by more active cognitive tasks (see later) and emotions. This is especially relevant since consciousness has often been associated with higher-order cognitive functions like imagination, memory, executive functions, attentions, and so on. Therefore, loss of consciousness in VS, for instance, was tacitly assumed to be associated with loss of cognitive functions, including their "willful modulation" by the subject itself (see Hohwy, 2012a, 2012b, for a nice overview of the different functions of consciousness in vegetative state; see focus here mainly on the purely phenomenal aspects).

Based on these findings, one may want to raise the following question: Is consciousness based on cognitive functions and thus cognition-based? I will first discuss various findings from recent studies in VS. This will lead me to reject the hypothesis that consciousness, that is, phenomenal consciousness, is based on cognitive function and thus cognition-based. Instead, consciousness is based on the phenomenal functions of the brain as they are related to its resting state activity (Northoff, 2014a, 2014b).

Neuronal findings Ia: cognitive tasks induce region-specific neural activity in the vegetative state

As we all know only too well, life is full of surprises. And why should that be different in the case of the brain? Let us turn, therefore, to Adrian Owen.

Adrian Owen is a researcher who is interested in consciousness; he especially focuses on the absence of consciousness in VS. Back in Cambridge, England, he investigated one patient with VS during different imagery tasks. This yielded some rather amazing results, as I will describe (see Owen et al., 2006).

What did Adrian Owen do? He scanned a VS patient in fMRI and let him perform specific cognitive tasks. While lying in the scanner, the VS patient was instructed to perform motor and visual imagery tasks (Owen et al., 2006): the patient was asked to imagine playing tennis. Surprisingly this yielded neural activity in the supplementary motor area in the VS patients. This region is related to movements as one imagines or executes them when playing tennis either mentally or physically. Most interestingly, the same region was activated in more or less the same way in healthy subjects. Hence, the VS patient was apparently able to perform a cognitive task as complex as imagining playing tennis. However, one cannot exclude that the observed neural activity is less based on the task itself but generated rather by pure chance.

To exclude such a possibility, Owen conducted the imaging during yet another task, a spatial navigation task, where the patient was asked to imagine visiting and walking around in the rooms of her house. As in the first task, neural activity changes were again induced—this time in other regions like the parahippocampal gyrus and the parietal cortex regions that are closely associated with spatial cognition as required by the task. The very same regions were also recruited in healthy subjects during the same task with regard to their own house or apartment.

Taken together, the results indicate that the VS patient was apparently quite able to perform a cognitive task like seeing visual and motor imagery. Most importantly, the VS patient was very able to differentiate between both tasks in the underlying neural activity patterns.

The results were recently replicated in a larger sample by Monti et al. (2010). Analogous paradigms were here conducted in a larger group of 54 patients, of whom 23 were diagnosed with VS and 31 with MCS (Monti et al., 2010). They had to perform the same tasks, imagining playing tennis and imagining walking from room to room in their own house. Five patients (four VS, one MCS) were indeed able to willfully modulate their neural activity during the tasks in a proper way: imagining playing tennis led to activation in the supplementary motor area (SMA) in all five patients, a region typically associated with either physical or imaginary movements.

In contrast, imagining walking in their own house induced neural activity changes in the parahippocampal gyrus in three VS and one MCS patients. These neural patterns were again similar to those in the healthy control subjects. Since then, other investigations of cognitive tasks requiring task-related efforts and willful modulation have been conducted in VS and MCS, with all showing some preserved neural activity in the respective regions in these patients (see Table 3 in Laureys and Schiff, 2012, for an overview).

Neuronal findings Ib: can the presence of consciousness be inferred from the presence of stimulus-induced (or task-related) activity?

What do these results tell us about VS in particular and consciousness in general? The presence of stimulus-induced activity lets many neuroscientists and philosophers propose that consciousness must be present, too. Otherwise, subjects would be unable to perform the cognitive tasks and elicit stimulus-induced activity. They thus infer the presence of consciousness from the presence of stimulus-induced and task-related activity.

Therefore, a subset of VS patients is these days described as showing "wakefulness," which is further specified as either "responsive" or "unresponsive" (Laureys and Schiff, 2012). However, other investigators have disputed and thus opposed this inference of the presence of consciousness from the observation of stimulus-induced and task-related activity in these patients (see Bernat, 2010; Hohwy, 2012a, 2012b; Monti et al., 2010; Nachev and Hacker, 2010; Panksepp et al., 2007, for discussion).

The opponents argue that the presence of a certain type of neuronal activity itself does not imply anything about the presence or absence of consciousness. Or, they put forward a more behavioral argument stating that the presence or absence of consciousness can only be decided on behavioral grounds, i.e., by the presence or absence of particular behavioral signs, rather than on purely neuronal grounds. We will not follow this discussion at this point in detail; we will come back to it, however, when discussing the relationship between cognition and consciousness in later sections.

Are the VS patients conscious? We do not know at this point, because the VS patients themselves are unable to tell us. What we do know for sure is that the VS patient investigated initially by Owen has regained consciousness since. And we know that these patients seem to show stimulus-induced or task-related activity. That is what we know at this point in time.

In contrast, we do not know whether such stimulus-induced activity that is purely neuronal by itself is accompanied by consciousness and its phenomenal features. More poignantly, we still do not know whether stimulus-induced or task-related activity necessarily or unavoidably entails its own association with consciousness and its phenomenal features.

Neuronal findings IIa: electrophysiological response to patient's own name in the vegetative state

We discussed so far how the brain in VS reacts to cognitive tasks. This, however, neglected self-specific stimuli seem to relate in a special way to the brain's intrinsic activity. The application of self-specific stimuli may therefore be of high interest in VS. How are self-specific stimuli like one's own name processed in the absence of consciousness and thus in VS?

One can present one's own name in an auditory way and record the related changes in neural activity by electrophysiological measures like electroencephalography (EEG). Do the VS patients show neural activity changes in response to their own names in the same way as they do during cognitive tasks as described earlier? A single case study investigated an MCS patient in EEG during stimulation with emotional stimuli (crying infant) and self-related stimuli (own name). They observed an almost "normal" activation pattern in the patient. The P300, a specific event-related component in EEG associated with cognitive processing, was well preserved while listening to especially the subject's own name (see Laureys et al., 2004).

A study by Perrin (Perrin et al., 2005; see also Perrin et al., 2006) observed the same during auditory evoked potentials in response to the subjects' own names in VS and MCS patients. The P300 was more or less preserved in all MCS patients and present in three of five VS patients. Only the onset or latency of the P300 was significantly delayed in MCS and VS patients compared to that in the healthy subjects.

Another study, by Schnakers (Schnakers et al., 2008), included 22 VS/MCS patients. Schnakers et al. demonstrated that subject's own name induced higher activity in another, later, more cognitive electrophysiological potential, the P300, compared to that in reaction to another person's name. This was stronger in an active (counting of names) than in a passive (mere perception without counting) mode. The difference between active and passive modes was observed only in MCS patients (14), while VS patients did not show any such difference. They were thus apparently unable to properly differentiate between the active and passive condition on a neuronal level.

Fellinger (Fellinger et al., 2011) also conducted an EEG study during one's own and unknown names that were presented in active and passive modes. Overall, the patients (13 MCS, 8 VS) showed stronger lower frequencies (delta, theta) and weaker higher frequencies (alpha, beta) than healthy subjects during hearing both their own and unknown names. Finally, frontal theta (at Fz) especially when hearing their own name was higher in the patients than the healthy subjects.

The pattern was different when the researchers compared active and passive modes of presentation. Healthy subjects showed stronger frontal theta power during the active mode compared to that in the passive mode. This was different in the patients. Like the earlier-mentioned study, the patients could not well differentiate between the two modes, i.e., active and passive, and also showed a delayed onset in frontal theta power compared to that in healthy subjects.

Neuronal findings IIb: preattentive processing of one's own name in the vegetative state

Probing another electrophysiological component in EEG, Pengmin Qin, from China, who is now in our group, investigated the same patients with EEG and focused on a specific electrophysiological potential, the MisMatch Negativity (MMN) (Qin et al., 2008).

The MMN taps into preattentive auditory sensory processing (at around 125–250 ms) by comparing the electrophysiological responses to the same repeating stimuli with the one during one deviant stimulus. To test for self-specificity in the MMN, Pengmin Qin determined the deviant stimulus as one's own name, while a non-self-specific name served as repeating stimulus.

The data show that Pengmin Qin's experimental design was well suited to eliciting an MMN during hearing their own name in all healthy subjects and in the seven patients (two coma, three VS, two MCS). Surprisingly there was no major difference in amplitude and latency in MMN between healthy subjects and the patients. In addition to the MMN, an earlier potential at around 100 ms (i.e., N100) could also be elicited in the seven patients and in two more patients. Interestingly, all the patients who reverted to MCS after three months showed an MMN and an N100. In contrast, no MMN (and N100) was observed in those VS patients who did not revert to MCS (see also Boly et al., 2011, for recent, more or less similar results on the MMN in VS).

What do these and other electrophysiological findings (see Cavinato et al., 2011, as well as Laureys and Schiff, 2012, for an overview of all studies) tell us about the stimulus-induced activity in VS and its relationship to consciousness? They demonstrate that self-specific stimuli can easily elicit neural activity changes in the brain of VS patients. The brain of these patients and thus their resting-state activity seem to be still reactive to stimuli like hearing one's own name. Accordingly, the electrophysiological results concerning self-specific stimuli are very compatible with the ones during cognitive tasks that, as described earlier, also showed preserved stimulus-induced activity in VS.

Neuronal findings IIc: neural activity in midline regions during self-specific stimuli predicts the degree of consciousness in the vegetative state

To investigate the functional anatomy, we turn from EEG and its electrophysiological measures to fMRI, which has a much better spatial resolution. The cortical midline regions seem to have a special role in processing self-specific stimuli. This raises the question of whether the VS patients and their midline regions' neural activity are still reactive to self-specific stimuli. There have indeed been two studies that tested for self-specificity in VS patients as conducted by our group.

Pengmin Qin from our group (Qin et al., 2010) auditorily presented one's own name to seven VS and four MCS patients while they were lying in the scanner (fMRI). He first mapped the relevant regions in healthy subjects by comparing one's own name to familiar and unfamiliar, that is, unknown, names. This yielded activity changes in various midline structures like the supragenual anterior cingulate cortex (sACC), dorsal anterior cingulate cortex (dACC), SMA, superior temporal gyrus (STG), posterior cingulate cortex (PCC), and bilateral insula.

What happens in these midline regions in VS and MCS during auditory presentation of one's own name? All patients were able to induce activity changes though to different degrees. The MCS patients showed higher neural activity in sACC, dACC, PCC, and SMA compared to that in the VS patients. This clearly suggests that these patients' midline regions are still somewhat reactive, meaning that they can induce neural activity changes during self-specific stimuli.

How is the midline activity during one's own name related to consciousness? Pengmin Qin observed significant correlation between the consciousness scores (as measured with the Coma Recovery Scale–Revised [CRS-R]) and the degree of neural activity in the dACC. The higher the signal change in the dACC during the auditory presentation of one's own name, the higher the degree of consciousness the patients exhibited. Those patients with VS showing the highest signal changes were the ones who were most likely to revert to MCS three months later.

One may now want to argue that one cannot be completely sure whether subjects really listened to their own name. The name was presented in a merely passive way requiring no active effort by the subjects to listen so that subjects may have simply not even listened to the name. One can therefore not exclude the neural activity change to stem from sources other than their own name. Hence, one would need an active task where subjects have to actively relate the stimulus to themselves, that is, their own self.

Neuronal findings IId: active self-referential task leads to decreased self–non-self differentiation of midline neural activity in the vegetative state

This is exactly what a subsequent fMRI study of ours in VS by Huang did (Huang et al., 2014). Instead of letting subjects merely passively listen to their own name, they now had to perform an active self-referential task wherein they had to refer to themselves, i.e., their own self. Two types of questions, autobiographical and common-sense, were presented in the auditory mode. The autobiographical questions asked for real facts in subjects' lives as obtained from their relatives.

This required subjects to actively refer the question to their own self, thus being a self-referential task. The control condition consisted of common-sense questions as non-self-referential, where subjects were asked for basic facts like whether one minute is 60 seconds. Instead of giving a real response via button click (as it is impossible in these patients), the subjects were asked to answer (mentally not behaviorally) with "yes" or "no."

Huang first compared autobiographical and common-sense questions in healthy subjects. As expected, this yielded significant signal changes in the midline regions, including the anterior regions like the perigenual anterior cingulate cortex (PACC) (extending to ventromedial prefrontal cortex [VMPFC]) and posterior regions like the PCC.

What did the brains in the VS patients now show in the very same regions? They showed signal changes in these regions that were reduced compared to those in healthy subjects. More specifically, while the VS patients were able to somehow differentiate between the two questions in their neural activity, the degree of neural differentiation remained much lower.

How are these signal changes now related to consciousness? As in the study by Pengmin Qin, a significant correlation in anterior midline regions was observed. The midline regions' activity, the PACC the dorsal anterior cingulate cortex (dACC), and the PCC correlated with the degree of consciousness (as measured with the CRS-R scale).

How is the exact relationship between neural activity in these regions and the level of consciousness? The better the signal changes in these regions differentiated neuronally between self- and non-self-referential conditions, the higher levels of consciousness patients exhibited. Accordingly, as in the earlier-described study, we here observed a direct relationship between the degree of neuronal self–non-self differentiation and the level of consciousness in anterior and posterior midline regions.

Neuronal findings IIe: resting-state activity in midline regions predicts stimulus-induced activity during self-referential task in the vegetative state

How about the resting-state activity in the same patients? Vegetative patients show strong alterations in the resting-state activity. One wants to know now whether the diminished responses to self-specific stimuli are related to changes in the resting state in the very same regions.

For that, Huang (Huang et al., 2014) also investigated functional connectivity and low-frequency fluctuations in exactly those regions that showed diminished signal differentiation during the self-referential task. The VS patients showed significantly reduced functional connectivity from the PACC to the PCC in the resting state. In addition, the power of particular ranges or bands in the low-frequency fluctuations was significantly lower in the PACC and the PCC in VS compared to that in healthy subjects.

Given that we investigated exactly the same regions during both resting state and task, this strongly suggests that the resting-state abnormalities in these regions are somehow related to the earlier described changes during the self-referential task. This was further supported by correlation analysis: The higher the degree of low-frequency fluctuations in the resting state of the midline regions, the better the stimulus-induced neuronal differentiation between self- and non-self-referential conditions.

Taken together, these findings demonstrate that VS patients not only can induce neural activity changes in their brain in response to merely passively presented self-specific stimuli. VS patients are apparently also able to actively refer to themselves and thus to engage by referring stimuli or questions to their own self, as required in the self-referential task. Thereby the anterior and posterior midline regions, like the anterior cingulate and its distinct parts (i.e., PACC, dACC, PCC), are recruited and seem to be of special significance for associated consciousness to the stimulus-induced or task-related activity.

Neuronal hypothesis Ia: "neuronal-phenomenal *dissociation*"

What do these findings tell us? First and foremost, they tell us that something must be "right" in the VS patients' brains. Otherwise they would not be able to induce neural activity changes during either cognitive or self-referential tasks. Nor would they be able to differentiate between the different tasks as, for instance, between motor (e.g., tennis playing) and visual (e.g., house navigation) imagery or between self- and non-self-referential stimuli.

These data suggest that what is "right" in VS concerns the induction of stimulus-induced and task-related activity and its relationship to specific tasks or stimuli. This was the easy part. Now comes the hard part. Something must also be "wrong" in the VS patients' brain. Even though they are quite able to induce stimulus-induced activity, they nevertheless seem to suffer from loss of consciousness, thus being vegetative.

More specifically, the stimulus-induced activity is apparently no longer associated with consciousness. There is thus what one may describe as a *dissociation* between stimulus-induced activity and consciousness. In contrast to the healthy brain, stimulus-induced activity in VS is no longer associated with consciousness. The purely neuronal stimulus-induced or task-related activity is thus dissociated from the phenomenal state of consciousness; one may therefore speak of "neuronal-phenomenal dissociation."

What exactly do I mean by the concept of "neuronal-phenomenal dissociation"? It means that neuronal and phenomenal states can no longer be characterized by co-occurrence. Even though there is neuronal activity like (more or less) proper stimulus-induced activity as in VS, it is no longer associated with a phenomenal state and thus consciousness. The stimulus-induced activity is consequently detached or dissociated from consciousness and its phenomenal features. This implies what I describe as neuronal-phenomenal dissociation (see later for a more detailed definition).

Neuronal hypothesis Ib: "neuronal-phenomenal *inference*"

One may now want to argue that such "neuronal-phenomenal dissociation" does not apply for those patients who are actively able to perform cognitive and self-referential tasks as described earlier. Does the presence of neuronal activity during the active cognitive and self-referential tasks signify the presence of consciousness? Such an inference, from the presence of stimulus-induced or task-related activity to the presence of consciousness, seems to be suggested by the most recent introduction of the terms "responsive" and "unresponsive wakefulness" to describe VS (see Laureys and Schiff, 2012).

What does the concept of "responsive and unresponsive wakefulness" mean? The terms "responsive" and "unresponsive" indicate whether these subjects show stimulus-induced or task-related activity in response to certain stimuli or tasks. The term "wakefulness" suggests the presence of an awake and somehow conscious state that is assumed to be necessary for performing the task. The presence of a phenomenal state; that is, consciousness as wakefulness, is here inferred from the presence of the purely neuronal stimulus-induced or task-related activity. Such inference from the presence of a neuronal state to the presence of consciousness and its phenomenal features can be described as the "neuronal-phenomenal inference."

Such a neuronal-phenomenal inference is problematic, however, for several reasons, both empirical and conceptual. Let us focus here on the empirical side of things (while I leave aside the conceptual-logical reasons). As Laureys and Schiff themselves remark (2012), the absence of neuronal activity in response to task-specific instructions may occur for several reasons (as, for instance, technological dependence). Therefore, the absence of neural activity cannot be taken as a marker for the absence of consciousness.

How about the reverse, the presence of task-specific neural activity indicating the presence of consciousness? Does task-specific neuronal activity require and thus presuppose consciousness? If so, these patients must be assumed to be conscious indeed and may therefore suffer from what Laureys and Schiff describe as "functional locked-in-syndrome" (2012). But one needs to be careful here.

Subjects remaining unconscious may show more or less the same activity pattern during the same kind of tasks. We perform plenty of tasks daily in a rather unconscious mode, meaning that we do not "experience" these tasks. We are thus both responsive and wakeful, but not conscious, with regard to these tasks. This means that responsiveness and wakefulness, including their underlying stimulus-induced or task-related activities, do not imply anything by themselves about the presence or absence of consciousness and its phenomenal features.

Neuronal hypothesis Ic: "neuronal-*phenomenal* inference" versus "neuronal-*cognitive* inference"

Let me be clear what exactly I mean here by the concept of "consciousness and its phenomenal features." The phenomenal features I am targeting here, are the "phenomenal features" in a strict sense, including "inner time and space consciousness," phenomenal unity, self-perspectival and intentional organization, and qualia. These phenomenal features must be distinguished from other, more cognitive features of consciousness like willful modulation, attention, awareness, and access to contents, which I do not debate here (see Hohwy, 2012a, 2012b, for an overview).

This implies a strict distinction between phenomenal and cognitive functions of the brain. The observed results with the neural activity during cognitive tasks suggest that the cognitive functions are somehow preserved in VS. One may thus reason from the presence of neural activity to the presence of the cognitive functions, making a so-called "neuronal-cognitive inference." That does not imply anything about the phenomenal functions themselves, however. To infer phenomenal features and consciousness from the observed neural activity is to confuse cognitive and phenomenal functions of the brain. Accordingly, the results allow for a "neuronal-cognitive inference" but not for a "neuronal-phenomenal inference."

Why do the proponents of the description of VS as "responsive or unresponsive wakefulness" nevertheless confuse these two inferences: the "neuronal-cognitive inference" and the "neuronal-phenomenal inference"? The tacit supposition here is that consciousness is based on cognitive functions and their associated stimulus-induced or task-related activity. This amounts to a cognition- and stimulus-based view of consciousness.

That, however, as we can see, does not really account for the data in VS. Here, the presence of stimulus-induced activity during cognitive tasks is accompanied by the absence of consciousness. This implies dissociation between the "neuronal-cognitive inference" and the "neuronal-phenomenal inference," with only the former, not the latter, being valid. Most important, the rejection of the "neuronal-phenomenal inference" forces us to develop a different account of consciousness, one that is not based on cognitive functions and stimulus-induced activity but rather on phenomenal functions and resting-state activity.

Neuronal hypothesis IIa: from "neuronal-phenomenal dissociation" to "neuronal-neuronal dissociation"

How about empirical reality? Empirical reality tells us that stimulus-induced and/or task-related activity is present in VS patients, while consciousness seems to be absent. How is such a dissociation between neuronal activity and phenomenal features possible? Let us briefly recapitulate what is clear and what is not in VS.

What is clear is that there is neuronal activity in VS and MCS patients in response to passive sensory stimuli and active cognitive tasks. That is a consistent finding, as described earlier. Their neural activity, the observed task-related activity, is still associated with particular psychological functions like imagining, navigation, self-referencing, and so on (see earlier). This suggests that there is apparently no dissociation between stimulus-induced activity and cognitive functions. There is no "neuronal-cognitive dissociation" in VS, as can be observed in depression or schizophrenia.

In addition, it is also clear that the VS patients show changes in their consciousness in that they are not able to properly associate their otherwise purely neuronal stimulus-induced or task-related activity with consciousness and its phenomenal features. They can no longer experience their own cognitive (and sensory, motor, affective, cognitive, and social) functions in a subjective way, in first-person perspective, as being indicative of consciousness. They thus show a phenomenal deficit, if one wants to say so. One may consequently postulate a dissociation between the neuronal activity during cognitive tasks and the phenomenal features of consciousness. As already indicated, I therefore speak of a "neuronal-phenomenal dissociation," in VS.

How can we further substantiate the concept of the "neuronal-phenomenal dissociation"? The observation of dissociation between two different states or functions usually implies that there must be two different underlying neuronal mechanisms. These two neuronal mechanisms may now dissociate from each other in VS, with one being intact and the other deficient.

What are the two neuronal mechanisms in question? There is the neuronal mechanism that enables the generation of stimulus-induced or task-related activity. And there is the neuronal mechanism that allows to associate the otherwise purely neuronal stimulus-induced or task-related activity with consciousness and its phenomenal features. What does this imply for the neuronal-phenomenal dissociation in VS? The neuronal-phenomenal dissociation suggests that the neuronal mechanisms for generating the neural activity during cognitive tasks are still more or less intact in VS. In contrast, the neuronal mechanisms related to the phenomenal features of consciousness seem to be deficient in VS.

The postulated neuronal-phenomenal dissociation in VS can be traced back to the dissociation between two different neuronal mechanisms: one for generating neural activity (during for instance cognitive functions), and the other for the association of that neural activity with the phenomenal features of consciousness. One can therefore specify the alleged "neuronal-phenomenal dissociation" by what I refer to as "neuronal-neuronal dissociation." What do I mean by "neuronal-neuronal dissociation"? This will be the focus in the next section.

Neuronal hypothesis IIb: from "neuronal-neuronal dissociation" to "rest–stimulus dissociation"

Exactly what kind of neuronal mechanism are we looking for? The neuronal mechanism in question must allow for the association of a phenomenal state with the purely neuronal stimulus-induced or task-related activity.

At the same time, however, the neuronal mechanism in question must be different from the ones underlying the generation of stimulus-induced or task-related activity by itself. Why? The neuronal mechanisms underlying the generation of the stimulus-induced or task-related activity by itself must be more or less preserved in VS, allowing them to show "normal" stimulus-induced activity. We must therefore search for a neuronal mechanism that lies beneath or beyond the stimulus-induced activity or task-related itself.

How can we better illustrate the situation? Metaphorically speaking, there must be an additional factor coming in besides the stimulus or task itself. And this additional factor must be crucial for associating the purely neuronal stimulus/task and its stimulus-induced or task-related activity with the phenomenal state of consciousness.

What is this additional factor? Let's look at what happens prior to the stimulus-induced activity. The stimulus must interact with the resting-state activity in order to elicit stimulus-induced activity. Such rest–stimulus interaction shows special features like nonlinear interaction via GABA-ergic-mediated neural inhibition, as we will see in further detail in the next section. I now propose that proper rest–stimulus interaction is central for associating the otherwise purely neuronal stimulus-induced or task-related activity with a phenomenal state and thus consciousness.

If, in contrast, rest–stimulus interaction is abnormal, that is, decreased, the resulting stimulus-induced or task-related activity will no longer be associated with consciousness anymore. There may thus be "neuronal-neuronal dissociation" between resting-state activity and stimulus-induced activity in VS.

Such neuronal dissociation may, in turn, be central for the loss of consciousness in VS. Since it concerns the coupling between resting-state and stimulus-induced activity, I describe such neuronal-neuronal dissociation also as "rest–stimulus dissociation."

Conclusion

Taken all together, I propose three different concepts of dissociation in VS. First, stimulus-induced or task-related activity dissociates from the phenomenal features of consciousness, implying what I describe as "neuronal-phenomenal dissociation." I trace such neuronal-phenomenal dissociation back to the decoupling of the neuronal mechanisms underlying stimulus-induced or task-related activity from those related to associating that neural activity with consciousness. I therefore spoke of "neuronal-neuronal dissociation."

I now postulate that the neuronal-neuronal dissociation can be traced back to the decoupling between resting-state activity and stimulus-induced activity. For that reason, I speak of "rest–stimulus dissociation"; this concept can be regarded as the empirical specification of the more general concepts of "neuronal-phenomenal dissociation" and "neuronal-neuronal dissociation." Rest-stimulus dissociation means that the intrinsic resting state activity and the extrinsic stimulus no longer properly interact with each other anymore leading to altered or deficient rest-stimulus interaction (Northoff, 2014a, 2014b; Northoff et al., 2010).

The notion of rest-stimulus dissociation entails that the question for the presence versus absence of consciousness can no longer be decided upon the presence or absence of stimulus-induced activity with the consecutive neuronal-cognitive inference. Instead, the question for the presence versus absence of consciousness in UWS/VS is delegated to the investigation of how the stimulus interacts with the resting state activity. Such empirical investigation of rest-stimulus interaction allows then for what can be described as neuronal-phenomenal inference on the conceptual level.

References

Bernat, J. L. 2010. Current controversies in states of chronic unconsciousness. *Neurology*, 75, S33–8.

Boly, M., Garrido, M. I., Gosseries, O., Bruno, M. A., Boveroux, P., Schnakers, C., Massimini, M., Litvak, V., Laureys, S., & Friston, K. 2011. Preserved feedforward but impaired top-down processes in the vegetative state. *Science*, 332, 858–62.

Cavinato, M., Volpato, C., Silvoni, S., Sacchetto, M., Merico, A., & Piccione, F. 2011. Event-related brain potential modulation in patients with severe brain damage. *Clin Neurophysiol*, 122, 719–24.

Fellinger, R., Klimesch, W., Schnakers, C., Perrin, F., Freunberger, R., Gruber, W., Laureys, S., & Schabus, M. 2011. Cognitive processes in disorders of consciousness as revealed by EEG time-frequency analyses. *Clin Neurophysiol*, 122, 2177–84.

Hohwy, J. 2012a. Neural correlates and causal mechanisms. *Conscious Cogn*, 21, 691–2.

Hohwy, J. 2012b. Preserved aspects of consciousness in disorders of consciousness: A review and conceptual analysis. *J Consciousness Stud*, 19, 3–4.

Huang, Z., Dai, R., Wu, X., Yang, Z., Liu, D., Hu, J., Gao, L., Tang, W., Mao, Y., Jin, Y., Wu, X., Liu, B., Zhang, Y., Lu, L., Laureys, S., Weng, X., & Northoff, G. 2014. The self and its resting state in consciousness: An investigation of the vegetative state. *Hum Brain Mapp*, 35, 1997–2008.

Laureys, S., Perrin, F., Faymonville, M. E., Schnakers, C., Boly, M., Bartsch, V., Majerus, S., Moonen, G., & Maquet, P. 2004. Cerebral processing in the minimally conscious state. *Neurology*, 63, 916–8.

Laureys, S., & Schiff, N. D. 2012. Coma and consciousness: Paradigms (re)framed by neuroimaging. *Neuroimage*, 61, 478–91.

Monti, M. M., Vanhaudenhuyse, A., Coleman, M. R., Boly, M., Pickard, J. D., Tshibanda, L., Owen, A. M., & Laureys, S. 2010. Willful modulation of brain activity in disorders of consciousness. *N Engl J Med*, 362, 579–89.

Nachev, P., & Hacker, P. 2010. Covert cognition in the persistent vegetative state. *Prog Neurobiol*, 91, 68–76.

Northoff, G. 2014a. *Unlocking the brain. Volume 1: Coding.* Oxford: Oxford University Press.

Northoff, G. 2014b. *Unlocking the brain. Volume II: Consciousness.* Oxford: Oxford University Press.

Northoff, G., Qin, P., & Nakao, T. 2010. Rest-stimulus interaction in the brain: A review. *Trends Neurosci*, 33, 277–84.

Owen, A. M., Coleman, M. R., Boly, M., Davis, M. H., Laureys, S., & Pickard, J. D. 2006. Detecting awareness in the vegetative state. *Science*, 313, 1402.

Panksepp, J., Fuchs, T., Garcia, V. A., & Lesiak, A. 2007. Does any aspect of mind survive brain damage that typically leads to a persistent vegetative state? Ethical considerations. *Philos Ethics Humanit Med*, 2, 32.

Perrin, F., Maquet, P., Peigneux, P., Ruby, P., Degueldre, C., Balteau, E., Del Fiore, G., Moonen, G., Luxen, A., & Laureys, S. 2005. Neural mechanisms involved in the detection of our first name: A combined ERPs and PET study. *Neuropsychologia*, 43, 12–19.

Perrin, F., Schnakers, C., Schabus, M., Degueldre, C., Goldman, S., Bredart, S., Faymonville, M. E., Lamy, M., Moonen, G., Luxen, A., Maquet, P., & Laureys, S. 2006. Brain response to one's own name in vegetative state, minimally conscious state, and locked-in syndrome. *Arch Neurol*, 63, 562–9.

Qin, P., Di, H., Liu, Y., Yu, S., Gong, Q., Duncan, N., Weng, X., Laureys, S., & Northoff, G. 2010. Anterior cingulate activity and the self in disorders of consciousness. *Hum Brain Mapp*, 31, 1993–2002.

Qin, P., Di, H., Yan, X., Yu, S., Yu, D., Laureys, S., & Weng, X. 2008. Mismatch negativity to the patient's own name in chronic disorders of consciousness. *Neurosci Lett*, 448, 24–8.

Schnakers, C., Perrin, F., Schabus, M., Majerus, S., Ledoux, D., Damas, P., Boly, M., Vanhaudenhuyse, A., Bruno, M. A., Moonen, G., & Laureys, S. 2008. Voluntary brain processing in disorders of consciousness. *Neurology*, 71, 1614–20.

Part III

8 Ethical and deontological issues in paediatric clinical studies

An analysis of documents from national and international institutions

Carlo Petrini

Foreword

The enrolment of minors in clinical studies is problematic, but necessary.

From an ethical point of view, the *problems* are manifold and significant. The two that call for particular attention are the procedures adopted for acquiring valid consent and the acceptability of risk.

The *need* for these studies is largely associated with the fact that it is generally both ethically and scientifically wrong to assume that knowledge acquired from studies with adults applies equally to children. The progress of diseases differs in children and in adults; many diseases typically affect children and have no equivalent in adults; children and adults differ physiologically and their reactions to drugs also vary. Briefly, children are not small versions of adults.

Regulations governing clinical trials usually refer to "minors" as a generic category of subjects for whom parental consent is required. But there are obvious differences between a newborn infant and an adolescent; when a child attains a certain level of awareness he or she may express consent or dissent, and although neither is legally binding it is widely accepted that the opinion expressed should be held in due consideration (Shaddy et al., 2010). In accordance with the rest of this book, this chapter refers specifically to "non-communicative" infants, focusing in particular on the issue of their involvement in clinical studies. It should be borne in mind that the category of "minors" is larger.

Definition of a "clinical study"

"Clinical trials" are a specific type of "clinical research" or "clinical studies".

A precise definition can be found in Regulation 536/2014 of the European Union, according to which "'Clinical study' means any investigation in relation to humans intended: (a) to discover or verify the clinical, pharmacological or other pharmacodynamic effects of one or more medicinal products; (b) to identify any adverse reactions to one or more medicinal products; or (c) to study the absorption, distribution, metabolism and excretion of one or more medicinal products; with the objective of ascertaining the safety and/or efficacy of those medicinal products". The same regulation defines "clinical trials" as particular types of "clinical studies": "'Clinical trial' means a clinical study which fulfils any of the following conditions: (a) the assignment of the subject to a particular therapeutic strategy is decided in advance and does not fall within normal clinical practice of the Member State concerned; (b) the decision to prescribe the investigational medicinal products is taken together with the decision to include the subject in the clinical study; or (c) diagnostic or monitoring procedures in addition to normal clinical practice are applied to the subjects" (European Parliament and Council of the European Union, 2014).

The term "clinical trial" describes the procedure most typically implemented in the so-called "interventional research". The most typical form of non-interventional research is observational research (Gill, 2004).

Reference documents

Documents adopted by eminent national, international, and supranational institutions are highly relevant for bioethics. These documents can be of different kinds: declarations, treaties, codes, conventions, and guidelines, to name but a few. Some are binding; others, while not binding, are essential as references. Several of these documents are recognised in international law as "soft law", i.e., as having the potential to prompt political authorities to adopt their recommendations as statutory regulations while not actually obliging them to do so.

All the documents recognise the need for special safeguards for minors. This particular need in regard to health matters is explicitly recognised in the Convention on the Rights of the Child. The Article 3.2 states that "States Parties undertake to ensure the child such protection and care as is necessary for his or her well-being, taking into account the rights and duties of his or her parents, legal guardians, or other individuals legally responsible for him or her, and, to this end, shall take all appropriate legislative and administrative measures". Article 24.1 adds that "States Parties recognize the right of the child to the enjoyment of the highest attainable standard of health and to facilities for the treatment of illness and rehabilitation of health. States Parties shall strive to ensure that no child is deprived of his or her right of access to such health care services" (United Nations, 1989).

The ethics of clinical trials are dealt with both in general documents that refer also to minors and in others that apply specifically to minors. The following paragraphs illustrate some examples of both types.

Documents not specifically focused on children

The ethical principles associated to research with human subjects were defined largely in the wake of the serious violations of human rights perpetrated during the Second World War, some of which involved children.

The "Nuremberg Code" (Nuremberg Military Tribunal, 1949) is considered the progenitor of later documents on the ethics of research with humans.

Because of the seriousness of the human rights violations of the preceding years, the drafters of the Code prohibited research involving children, stating that "The voluntary consent of the human subject is absolutely essential. This means that the person involved should have legal capacity to give consent" (Article 1). Later documents gradually moved away from this total ban on trials involving persons unable to express consent and introduced criteria to allow such trials, albeit with strict safeguards. The exclusion of children (as of other categories of so-called "vulnerable" subjects) from trials protects them from the risks associated with such research, but at the same time excludes them from the possible benefits.

The so-called "Belmont Report" (The National Commission for the Protection of Human Subjects of Biomedical and Behavioral Research, 1979) was an early example of a less strict approach. It encoded four principles that became known as the "principles of biomedical ethics": respect for persons (or autonomy), beneficence, nonmaleficence, and justice. These principles were elaborated specifically with reference to research with humans but were rapidly extended to a much wider field and applied to any problems of biomedical ethics (Beauchamp and Childress, 2012).

The principle of respect for persons (or autonomy) aims to ensure that individuals are treated as autonomous agents and that subjects with diminished autonomy are given protection. In the setting of paediatric research, "most research subjects are not capable of making autonomous decisions because they do not comply with the ethical and legal requirements to do so". The fact that most minors are unable to give legally valid consent, however, does not mean that they are without some ability to make decisions, nor that they are unable to understand what a decision is about, to evaluate information, or to make rational decisions (Pinxten et al., 2009).

The principle of beneficence means that human subjects should not be harmed and that research should maximise the potential benefits and minimise any possible harmful effects. In the context of paediatrics this principle is manifest in efforts to offset the risks and burdens associated with participation in research (Pinxten et al., 2009). The benefit to the single participant and the advancement of science for the benefit of the community as a whole are frequently in conflict and all the major documents on the ethics of biomedical research acknowledge that the former should prevail over the latter. Article 8 of the World Medical Association's Declaration of Helsinki, for instance, states that "While the primary purpose of medical research is to generate new knowledge, this goal can never take precedence over the rights and interests of individual research subjects" (World Medical Association, 1964–2013).

As regards nonmaleficence in paediatric research, the difficulty of balancing the protection of minors and the advancement of medical science means that the aim of preventing unethical research may hinder or even stop the development of new drugs for paediatric use. Therefore, "paediatric expertise in ethics committees is essential in addressing the specific complexities of involving minors in clinical trials" (Pinxten et al., 2009).

With regard to justice, the procedures for obtaining authorisation to market drugs ensure that drugs entering the market are safe and effective. Anyway, these procedures are not able to ensure a sufficient variety of drugs suitable for minors. The extremely complex procedures for testing drugs in children imply that the costs may be higher than the potential return on the manufacturer's investment, thereby reducing the economic incentive to label drugs for paediatric use, which in turn leads to a lack of drugs specifically for children. "In many instances, pediatricians have no therapeutic options apart from using drugs off-license or off-label" (Pinxten et al., 2009).

The Council of Europe's "Convention for the Protection of Human Rights and Dignity of the Human Being with regard to the Application of Biology and Medicine: Convention on Human Rights and Biomedicine" was signed at Oviedo on 4 April 1997 (Council of Europe, 1997a). The convention, which was accompanied by an Explanatory Report (Council of Europe, 1997b), was followed by a number of Additional Protocols, one of which concerns biomedical research (Council of Europe, 2005).

Articles 16 and 17 are particularly relevant for research involving children: they define the criteria for authorising clinical studies in general as well as those involving persons unable to give valid consent. The criteria are very clear and it is worth fully quoting them. Article 16, covering "Protection of persons undergoing research", states that "Research on a person may only be undertaken if all the following conditions are met: i) there is no alternative of comparable effectiveness to research on humans; ii) the risks which may be incurred by that person are not disproportionate to the potential benefits of the research; iii) the research project has been approved by the competent body after independent examination (…); iv) the persons undergoing research have been informed of their rights and the safeguards prescribed by law for their protection; v) the necessary consent (…) has been given expressly, specifically, and is documented. Such consent may be freely withdrawn at any time". Article 17, concerning "Protection of persons not able to consent to research", states that "Research on a person without the capacity to consent (…) may be undertaken only if all the following conditions are met: i) the conditions laid down in Article 16, sub-paragraphs i to iv, are fulfilled; ii) the results of the research have the potential to produce real and direct benefit to his or her health; iii) research of comparable effectiveness cannot be carried out on individuals capable of giving consent; iv) the necessary authorisation (…) has been given specifically and in writing; v) and the person concerned does not object" (Council of Europe, 1997a).

Documents specifically focusing on research with minors

There is also a large number of documents that specifically address the issue of research with minors. One of the more exhaustive is a broad-ranging text published by the Institute of Medicine Committee on Clinical Research Involving Children (Field and Berman, 2004).

These documents address many of the ethical problems associated with research with minors. According to the American Academy of Pediatrics (Shaddy et al., 2010), the ethical issues of particular concern in drug investigation in paediatric research are determination of benefit and risks; evaluation by data- and safety-monitoring committees; informed permission/consent/assent; the permission process; waiver of permission; emergency research; permission for studies with life-threatening illness; institutionalized children; and assent of the child.

The American Academy of Pediatrics states that proposals for investigation of drugs involving children must include measures to protect the interests of children and must be scientifically sound and significant; be directed by investigators who operate in a state of scientific uncertainty; include a robust plan to monitor safety; consider the unique physiology, anatomy, psychology, pharmacology, social situation, and special needs of children; minimise risk while maximising benefit; take into account the racial, ethnic, gender, and socioeconomic characteristics of children; and conform to all relevant laws and guidelines (Shaddy et al., 2010).

Another US institution, the Presidential Commission for the Study of Bioethical Issues (Presidential Commission for the Study of Bioethical Issues, 2013), defined a set of conditions to be satisfied in order to ensure that the research complied with "sound ethical principles". These conditions are divided into five categories: an ethically acceptable risk threshold and adequate protection from harm; an ethical study and trial design; post-trial requirements to ensure ethical treatment of children and their families; community engagement; and transparency and accountability. The commission also reiterated the importance of informed parental permission and meaningful and developmentally appropriate assent by children.

Another relevant document is the "Guidelines for the ethical conduct of medical research involving children" published by the Royal College of Paediatrics and Child Health (Royal College of Paediatrics and Child Health, 2000). This defines six reference principles: research that involves children is an important source of benefit for all children and should be supported, encouraged, and conducted in an ethical manner; the interests of children are unique to them and they should not be considered as small adults; children should become involved in research only if no comparable studies involving adults can provide answers to the same questions; research procedures that are not directly designed for the benefit of the child subject are not *ipso facto* either unethical or illegal; all research projects that propose to involve children should be assessed by a research ethics committee; and legally valid consent should be obtained from the child, parent, or guardian, as appropriate. In the case of parental consent, researchers should also seek the agreement of any school-age children involved in the study.

Other relevant documents are "Medical Research Involving Children" published by the British Medical Research Council (Medical Research Council, 2004); the document by the Swiss National Advisory Commission on Biomedical Ethics (Swiss National Advisory Commission on Biomedical Ethics, 2009); that by various Canadian institutions (McGill University Centre for Genomics and

Policy et al., 2012); and the Standards for Research in (StaR) Child Health (Hartling et al., 2011).

These documents essentially share a similar approach to the basic ethical principles involved in research with minors, while they differ mainly on operational aspects. Among the imperative requisites stated by all these documents are the following:

- The interests of minors involved in research procedures must prevail over those of science and the community at large;
- There must be an acceptable balance between risks and benefits;
- Consent must be obtained from whoever exercises parental authority;
- The research must be authorised by an ethics committee; and
- Children should be involved only when the relevant knowledge cannot be obtained through research with adults.

Regulations relevant for research with minors

This section introduces some of the most significant international regulations about research involving minors. The regulation explicitly addresses also ethical issues.

Article 32 of the European Union Regulation 53/2014 establishes the requisites for obtaining informed consent for clinical trials involving minors. As well as the general requisites regarding informed consent, other more specific conditions must be met:

(a) the informed consent of their legally designated representative has been obtained;

(b) the minors have received the information (…) in a way adapted to their age and mental maturity and from investigators or members of the investigating team who are trained or experienced in working with children;

(c) the explicit wish of a minor who is capable of forming an opinion and assessing the information (…) is respected by the investigator;

(d) no incentives or financial inducements are given (…);

(e) the clinical trial is intended to investigate treatments for a medical condition that only occurs in minors or the clinical trial is essential with respect to minors to validate data obtained in clinical trials on persons able to give informed consent or by other research methods;

(f) the clinical trial either relates directly to a medical condition from which the minor concerned suffers or is of such a nature that it can only be carried out on minors;

(g) there are scientific grounds for expecting that participation in the clinical trial will produce:

  (i) a direct benefit for the minor concerned outweighing the risks and burdens involved; or

(ii) some benefit for the population represented by the minor concerned and such a clinical trial will pose only minimal risk to, and will impose minimal burden on, the minor concerned in comparison with the standard treatment of the minor's condition. (European Parliament and Council of the European Union, 2014)

In addition: "The minor shall take part in the informed consent procedure in a way adapted to his or her age and mental maturity" (European Parliament and Council of the European Union, 2014).

Another fundamental reference for trials involving minors in the European Union is Regulation 1901/2006 (European Parliament and Council of the European Union, 2006a) subsequently amended by Regulation 1902/2006 (European Parliament and Council of the European Union, 2006b). This not only establishes the obligation to conduct clinical trials with minors for every new drug developed for adults that can potentially be used for children, but also provides for a series of incentives for paediatric trials; the creation of a special fund (Medicines Investigation for the Children of Europe) and a European network of specialised facilities; requisites of the Paediatric Investigation Plan; and the establishment of the Paediatric Committee within the European Medicines Agency (European Commission, 2013).

In the USA the reference institution is the Food and Drug Administration (FDA), with its regulation of paediatric research contained in the Code of Federal Regulations (CFR) (Title 21 CFR 50-50-50.55: Subpart D: Additional Safeguards for Children in Clinical Investigations) (Code of Federal Regulations, 2009a). The provisions correspond to those set out by the Department of Health and Human Services (DHS), which regulates paediatric trials financed with federal funds, in 45 CFR 46 subpart D (Code of Federal Regulations, 2009b).

Another key international reference is the International Conference on Harmonisation of Technical Requirements for Registration of Pharmaceuticals for Human Use (ICH) Guideline E11 2.6, which specifically concerns paediatric trials (International Conference on Harmonisation of Technical Requirements for Registration of Pharmaceuticals for Human Use, 2000). The ICH guideline forms a common ground for Europe, Japan, and the US.

Two crucial elements: valid consent and the risk/benefit ratio

Several of the ethical problems associated with trials involving minors are similarly arising from studies with other categories of so-called "vulnerable" subjects.

Many of these ethical problems can be condensed in two key questions: "Who should make a specific decision?" and "Which decision should be made?" The former is a matter of procedure and specifically concerns consent; the latter is a matter of substance and concerns the risk/benefit ratio.

Consent

Consent embodies the principle of autonomy (Beauchamp and Childress, 2012), which in turn is one of the cornerstones of bioethics.

When a patient or a research subject is not able to express valid consent, as is the case of minors and adults suffering from partial or total incapacity, consent must be obtained by proxy (Mazur, 2012). In these cases the regulations require that a legal representative must give consent. This requirement for a legal representative is clearly set out in regulations (European Parliament and Council of the European Union, 2014, Article 28.1b), as well as in the key documents dealing with research ethics. The "Convention on Human Rights and Biomedicine", for instance, establishes that "Where, according to law, a minor does not have the capacity to consent to an intervention, the intervention may only be carried out with the authorisation of his or her representative or an authority of a person or body provided for by law" (Council of Europe, 1997a, Article 6.2). In the case of minors, consent is given by whoever exercises parental authority or, if no such person exists, by the legally appointed representative.

All the main relevant institutions and reference documents stress that, while consent expressed by the person who exercises parental authority or by a legal representative is the only consent that is legally valid until the minor reaches his or her majority, there is a duty not only to try to know the wishes of minors who have attained an appropriate ability to comprehend and to express their opinions consciously, but to take carefully into consideration their assent or dissent when decisions are being made. Baines even suggests that "an approach which relies on the consent of competent children and parental consent for incompetent children is to be preferred" (Baines, 2011).

While there is almost general agreement that the minor's wishes should be taken into consideration, the opinions concerning how these should be expressed differ widely, as do those regarding the criteria used to determine whether or not a particular minor is able to express his or her wishes with sufficient competence. The sole age of the child is clearly not an appropriate criterion: individual characteristics are such that it is not enough to establish a particular age at which a minor should be considered sufficiently mature (Baines, 2011).

Obviously, in the case of infants in the first few years of life, who are unable to communicate, the question of assent does not arise directly. Anyway, the increasing development of tools and methods to explore the brain of infants gives us new opportunities to visualise their cerebral life and make appropriate inferences concerning their mental life development (Dubois et al, 2012; Shukla and Ciaramitaro, 2016). This growing collection of data give us new opportunities to infer relevant information about the brain and mental state of non-communicative subjects, e.g., infants, for making decisions concerning their health assistance that are more in accordance with their actual needs. In this way, neurotechnology potentially makes available new tools for an informed proxy decision by adults. Moreover, in the case of children, neurotechnology could give us new opportunities to assess their competence for eventual involvement in the decision-making process.

Without questioning the validity of and the need for consent given by a legal representative, it cannot be ignored that the very notion of "proxy consent" is a misnomer, "precisely because the potential subjects are noncompetent, 'voiceless' persons incapable of giving consent". Thus the given consent is the "personal consent of parents and other guardians" rather than that of their wards (May, 2007). Neurotechnology could be relevant in order to include the needs of speechless infants in order to make the proxy decisions closer to the patients' needs.

Risk/benefit ratio

When assessing the risks and benefits associated with clinical studies involving minors there is one fundamental ethical principle that must be considered. It is expressed in Article 8 of the Declaration of Helsinki: "While the primary purpose of medical research is to generate new knowledge, this goal can never take precedence over the rights and interests of individual research subjects" (World Medical Association, 1964–2013). The benefit that the research may bring to the community should not be overlooked, but it cannot take precedence over the individual good of the person taking part in the trial. We may feel somewhat perplexed on reading the position adopted by the International Bioethics Committee of UNESCO, according to whom, "Research activities involving children are carried out to learn more about the nature of paediatric development, disease and potential treatments. Though one might hope that it will in some cases be beneficial to the research participant, the activity cannot be said to be specifically designed for this purpose because of the nature of the research question" (UNESCO, 2008). From an ethical point of view, to give priority to the good of the community over that of the individual would nonetheless be a betrayal of the Kantian imperative ("Act in such a way that you treat humanity [...] as an end and never merely as a means to an end") (Kant, 1998).

Having established that the interest of the patient participating in a research study should take precedence over that of science (except in very rare circumstances), a number of questions arise, two of which deserve special consideration. The first is the procedure for determining the patient's "best interest"; the second concerns the acceptability of risk and the assessment of the so-called "minimal risk".

At the crossroads between the issues of informed consent and risk lies the notion of the "best interest" of the child taking part in research. This notion is invoked for children who are not able to express their preferences. The concept of "best interest" represents an approach that "promotes an effort to be more objective, weighing the potential benefits and burdens for a particular child" (Kodish, 2005).

In the context of paediatric research, the notion of "best interest" refers mainly to the principle of beneficence (Beauchamp and Childress, 2012). While in the ethics of research with adults autonomy takes priority over beneficence, the ethics of paediatrics generally accord greater importance to beneficence than to autonomy.

The concept of best interest requires that the parent or legal representative should decide what, in his or her view, "would benefit the patient most" (Berg et al., 2001). It acknowledges the fact that the decision is based not on the subjective preferences of whoever is legally authorised to decide, but rather on "the highest net benefit among the available options, assigning different weights to interests the patient has in each option and discounting or subtracting inherent risks or costs" (Beauchamp and Childress, 2012).

The "best interest" concept thus presumes that the patient's interest lies in that which is of benefit to him or her, and that that benefit lies in whatever contributes to his or her well-being. This raises a number of problems for those who are required to take decisions, since well-being is a multidimensional concept. One kind of well-being refers exclusively to a person's health, while another is defined by Veatch as "total" and as including the "entirety" of the patient's well-being (Veatch, 2003). However, even if we refer only to the medical aspect of well-being it is not easy to determine what constitutes a "benefit". According to Veatch it comprises at least four indispensable elements: preventing death, curing diseases, alleviating pain, and promoting the well-being of the patient (Veatch, 2003, p. 53). Other lists of the basic ingredients of a child's interests range beyond the merely clinical aspects. The list drawn up by Malek, for example, comprises 13 elements (Malek, 2009).

The assessment of the above-mentioned elements is not an easy task, especially in the case of speechless subjects like infants. Here again neurotechnology, particularly neuroimaging technologies, could potentially allow a more direct detection of cerebral and mental signatures of diseases, pain, and well-being, i.e., of the best interest.

In the perspective of research ethics, a strict interpretation of "best interest" might preclude any type of research that does not offer a direct benefit to the subject concerned. Nonetheless, as already noted, we cannot deny that research is also in the interest of the community, which benefits from the advancement of knowledge, and not only of the individual directly involved in research.

The "best interest" approach has positive aspects, including the fact that it allows cases to be assessed individually; it reminds the physician of his responsibility for his decisions involving minors and his position of guarantor in the event his views conflict with the parents' expectations; and it focuses attention not on the decision-maker but on the minor. At the same time the "best interest" concept raises problems: it is simplistic and may be difficult to apply in complicated cases; because it is based mainly on a comparison between risks and benefits, it is pragmatic-utilitarian in outlook (Lyon, 2001); and it is not easily applied to those not involved in any way in the decision-making process.

Because the "best interest" notion can be difficult to apply, it has been suggested to be replaced with the notion of "avoidance of harm" (Miller and Brody, 2003), which, however, is too strongly focused on protecting the minor from risks and harm, whereas the "best interest" approach is more appropriately aimed at promoting his or her good.

The regulations of the US define "minimal risk" as meaning "that the probability and magnitude of harm or discomfort anticipated in the research are not greater in and of themselves than those ordinarily encountered in daily life or during the performance of routine physical or psychological examinations or tests" (Code of Federal Regulations, 2009c).

The Institutional Review Boards may also approve studies that pose a risk not greater than "minimal" (or with a minimal or minor increase over minimal) even where no direct benefit is envisaged (Code of Federal Regulations, 2009d). Studies that carry a higher risk may be approved only if there is a "prospect of direct benefit".

Defining a threshold of acceptability nonetheless remains a complex issue.

As an example, many children are regularly passengers in road vehicles, a circumstance that poses "the highest risk of mortality ordinarily encountered by healthy children". This risk ranges from six deaths per 100 million car trips for children aged 14 years and younger to 40 per 100 million for children aged 15 to 19 years (Wendler et al., 2005). Clearly it would be unacceptable to countenance a similar probability of risk in a trial not expected to bring direct benefit to the participants (Schmidt et al., 2011), but the situation is obviously different if such a benefit is expected. It may, for example, be legitimate to include a gravely ill child in a risky trial if there are well-founded hopes that he or she will derive benefit, though even in this case the definitions of "minimal risk" may be controversial. What risk, for instance, is to be expected "during the performance of routine physical or physiological tests"? For a healthy child a "routine" test may be giving a blood sample, but sick children may necessarily be "routinely" subjected to invasive and risky tests.

In the particular case of neurological diseases, brain modelling could be an important tool to improve the diagnosis and even the prognosis of the patients, namely of infants and minors in general, assessing prospective risks and benefits. Anyway, to date this perspective seems to be more futuristic than actual: the large-scale or the whole-scale simulation of the brain are still in their infancy, and more theoretical and technical efforts are needed in order to achieve the goal of an informative and predictive simulation of the brain and its translation in clinical context.

We thus return to the already mentioned question of weighing the direct benefit to the minor participating in a study (which must prevail) and the interest of the community derived from the advancement of scientific knowledge. According to Wendler, where the protection of subjects who cannot give their consent is concerned, "We need a new standard for what 'minimal' risks are, however – one that recognizes that participating in nonbeneficial research is like participating in a charitable activity" (Wendler, 2005), and the author recognises that any such standard will probably lead to more stringent protection for these vulnerable subjects (Wendler, 2005). However high the value we attribute to altruism and philanthropy, when the decisions concern vulnerable subjects who are not able to give consent in person, the greatest care must be taken to ensure that they are not exposed to risks without receiving a direct benefit. Neurotechnological advancement in the direction to directly assess cerebral and mental states of speechless patients is a potential new way for balancing risk and benefit and promoting their well-being. Anyway, more scientific and ethical reflections are needed concerning both the feasibility and the implementation of such hypothetical cerebral communication.

Acknowledgment

The author is very grateful to Michele Farisco, who provided very useful suggestions on a first draft of this chapter.

References

Baines, P. 2011. Assent for children's participation in research is incoherent and wrong. *Arch Dis Child*, 99(10), 960–2.

Beauchamp T, & Childress J. F. 2012. 7th ed. (1st ed.: 1978). *Principles of biomedical ethics*. Oxford and New York: Oxford University Press.

Berg, J. W., Appelbaum, P. S., Lidz, C. W., & Parker, L. S. 2001. (2nd ed.). *Informed consent: Legal theory and clinical practice*. Oxford and New York: Oxford University Press.

Code of Federal Regulations. 2009a. Federal Policy for the Protection of Human Research Subjects (Common Rule) a. 21 CFR (Code of Federal Regulations) 50.50–50.55.

Code of Federal Regulations. 2009b. Federal Policy for the Protection of Human Research Subjects (Common Rule) b. 45 CFR (Code of Federal Regulations) 46, subpart D.

Code of Federal Regulations. 2009c. Federal Policy for the Protection of Human Research Subjects (Common Rule) b. 45 CFR (Code of Federal Regulations) 46, 102(i).

Code of Federal Regulations. 2009d. Federal Policy for the Protection of Human Research Subjects (Common Rule) b. 45 CFR (Code of Federal Regulations) 46, 404.

Council of Europe. 1997a. Convention for the Protection of Human Rights and Dignity of the Human Being with regard to the Application of Biology and Medicine: Convention on Human Rights and Biomedicine. Accessed 12 March 2015 from http://conventions.coe.int/Treaty/en/Treaties/html/164.htm.

Council of Europe. 1997b. Convention for the Protection of Human Rights and Dignity of the Human Being with regard to the Application of Biology and Medicine: Convention on Human Rights and Biomedicine. Explanatory Report. Accessed 12 March 2015 from http://conventions.coe.int/Treaty/EN/Reports/Html/164.htm.

Council of Europe. 2005. Additional protocol on the Convention on Human Rights and Biomedicine concerning Biomedical Research. Accessed 12 March 2015 from http://conventions.coe.int/Treaty/en/Treaties/html/195.htm.

Dubois, J., Dehaene-Lambertz, G., Mangin, J. F., Le Bihan, D., Hüppi, P. S., & Hertz-Pannier, L. 2012. Brain development of infant and MRI by diffusion tensor imaging. *Neuropysiol Clin*, 42(1–2), 1–9.

European Commission. 2013. Report from the Commission to the European Parliament and the Council. Better Medicines for Children – From Concept to Reality. General Report on experience acquired as a result of the application of Regulation (EC) No 1901/2006 on medicinal products for paediatric use.

COM(2013) 443. Accessed 12 March 2015 from http://ec.europa.eu/health/files/paediatrics/2013_com443/paediatric_report-com(2013)443_en.pdf.

European Parliament & Council of the European Union. 2006a. Regulation (EC) No 1901/2006 of the European Parliament and of the Council of 12 December 2006 on medicinal products for paediatric use and amending Regulation (EEC) No 1768/92, Directive 2001/20/EC, Directive 2001/83/EC and Regulation (EC) No 726/2004. *Official Journal of the European Union*, L378, 1–19.

European Parliament & Council of the European Union. 2006b. Regulation (EC) No 1902/2006 of the European Parliament and of the Council of 20 December 2006 amending Regulation 1901/2006 on medicinal products for paediatric use. *Official Journal of the European Union* 27, L378, 20–1.

European Parliament & Council of the European Union. 2014. Regulation (EU) No 536/2014 of the European Parliament and of the Council of 16 April 2014 on clinical trials on medicinal products for human use, and repealing Directive 2001/20/EC. *Official Journal of the European Union*, L158, 1–76.

Field, M. J., & Berman, R. E. (eds.). 2004. Committee on Clinical Research Involving Children, Institute of Medicine. The ethical conduct of clinical research involving children. Washington, D.C.: The National Academies Press.

Gill, D. 2004. Ethical principles and operational guidelines for good clinical practice in paediatric research. Recommendations of the Ethics Working Group of the Confederation of European Specialists in Paediatrics (CESP). *Eur J Pediatr*, 163(2), 53–7.

Hartling, L., Wittmeier, K. D. M., Caldwell, P. H., Van Der Lee, J. H., Klassen, T. P., Craig, J. C., & Offringa, M. 2011. StaR child health: Developing evidence-based guidance for the design, conduct, and reporting of pediatric trials. *Clin Pharmacol Ther*, 90(5), 727–31.

International Conference on Harmonisation of Technical Requirements for Registration of Pharmaceuticals for Human Use – Ich. 2000. Guidance for Industry. E11 Clinical Investigation of Medicinal Products in the Pediatric Population. Accessed 12 March 2015 from www.fda.gov/downloads/RegulatoryInformation/Guidances/ucm129477.pdf.

Kant, I. 1998. (1st ed.: 1785). *Groundwork of the metaphysic of morals.* (Transl. Mark Gregor). Cambridge: Cambridge University Press.

Kodish, E. 2005. Ethics and research with children: an introduction. In KODISH, E. (ed.), *Ethics and research with children. A case-based approach*, pp. 3–21. Oxford: Oxford University Press.

Levine, R. J. 2008. The nature, scope, and justification of clinical research. In Emanuel, E. J., Grady, C., Crouch, R. A., Lier, R. K., Miller, F. G., & Wendler, D. (eds.), *The Oxford textbook of clinical research ethics*, pp. 211–21. Oxford: Oxford University Press.

Lyon, D. 2001. Utilitarianism. In Becker, L., & Becker, C. (eds.), *Encyclopedia of ethics*, Vol. 3, pp. 1737–44. New York: Routledge.

Malek, J. 2009. What really is in a child's best interest? Toward a more precise picture of the interests of children. *J Clin Ethics*, 20(2), 17–825.

May, W. E. 2007. Proxy consent for nontherapeutic experimentation. *Natl Cathol Bioeth Q*, 7(2), 239–47.

Mazur, G. 2012. *Informed consent, proxy consent, and Catholic bioethics.* Dordrecht: Springer.

McGill University Centre for Genomics and Policy, Maternal Infant Child and Youth Research Network, National Council on Ethics in Human Research, Avard, D., Black, L., Samuël, J., & Knoppers, B. M. (eds.). 2012. Best practices for research involving children and adolescents. Genetic, pharmaceutical and longitudinal studies. Accessed 12 March 2015 from www.genomicsandpolicy.org/en/best-practices-2012.

Medical Research Council. 2004. Medical research involving children. Accessed 12 March 2015 from www.mrc.ac.uk/documents/pdf/medical-research-involving-children/.

Miller, F. G., & Brody, H. 2003. A critique of clinical equipoise. Therapeutic misconception in the ethics of clinical trials. *Hastings Cent Rep*, 33(3), 19–28.

The National Commission for the Protection of Human Subjects of Biomedical and Behavioral Research. 1979. The Belmont Report: Ethical Principles and Guidelines for the Protection of Human Subjects of Research. Publication OS 78-0012. Accessed 12 March 2015 from www.hhs.gov/ohrp/humansubjects/guidance/belmont.html.

Nuremberg Military Tribunal. 1949. Permissible medical experiment [known as The Nuremberg Code]. In Trials of war criminals before the Nuremberg Military Tribunals under Control Council Law n. 10. Nuremberg, October 1946–April 1949. Washington, DC: Government Printing Office, Vol. 2, 181–2.

Pinxten, W., Dierickx, K., & Nys, H. 2009. Ethical principles and legal requirements for pediatric research in the EU: An analysis of the European normative and legal framework surrounding pediatric clinical trials. *Eur J Pediatr*, 168(10), 1225–34.

Presidential Commission for the Study of Bioethical Issues. 2013. Safeguarding Children: Pediatric Medical Countermeasure Research. Accessed 12 March 2015 from http://bioethics.gov/sites/default/files/PCSBI_Pediatric-MCM508.pdf.

Royal College of Paediatrics and Child Health. 2000. Guidelines for the ethical conduct of medical research involving children. *Arch Dis Child*, 82(2), 177–82.

Schmidt, M. H., Marshall, J., Downie, J., & Hadskis, M. R. 2011. Pediatric magnetic resonance research and the minimal-risk standard. *IRB*, 33(5), 1–6.

Shaddy, R. E., Denne, S. C., & The Committee on Drugs and Committee on Pediatric Research. 2010. Guidelines for the ethical conduct of studies to evaluate drugs in pediatric populations. *Pediatrics*, 125(4), 850–60.

Shukla, M., & Ciaramitaro, V. 2016. Cognitive capacities of the infant mind – a neuroimaging perspective. In Farisco, M., & Evers, K., *Neurotechnology and direct brain communication. New insights and responsibilities concerning speechless but communicative subjects.* Oxford: Routledge.

Swiss National Advisory Commission on Biomedical Ethics. 2009. Opinion n. 16. Research involving children. Accessed 12 March 2015 from www.nek-cne.ch/fileadmin/nek-cne-dateien/Themen/Stellungnahmen/en/NEK-CNE_Research_involving_children.pdf.

Unesco – United Nations Educational, Scientific and Cultural Organization. 2008. Report of the International Bioethics Committee of UNESCO (IBC) on Consent. Accessed 12 March 2015 from http://unesdoc.unesco.org/images/0017/001781/178124E.pdf.

United Nations. 1989. Convention on the Rights of the Child. Adopted and opened for signature, ratification, and accession by General Assembly resolution 44/25 of 20 November 1989 entry into force 2 September 1990, in accordance with article 49.

Veatch, R. 2003. *The basics of bioethics*. Upper Saddle River, NJ: Prentice Hall.

Wendler, D. 2005. Protecting subjects who cannot give consent. *Hastings Cent Rep*, 35(5), 37–43.

Wendler, D., Belsky, L., Thompson, K. M., & Emanuel, E. J. 2005. Quantifying the federal minimal risk standard: Implications for pediatric research without a prospect for direct benefit. *JAMA*, 294(7), 826–32.

World Medical Association. 1964–2013. Declaration of Helsinki – Ethical Principles for Medical Research Involving Human Subjects. Accessed 12 March 2015 from www.wma.net/en/30publications/10policies/b3/.

9 Disorders of consciousness and informed consent

Ralf J. Jox

In recent years, remarkable advances in neuroscience have indicated that some of the patients with disorders of consciousness (DOC) might have more residual awareness than clinically suspected (Owen et al., 2006; Monti et al., 2010; Cruse et al., 2011) and that these residually conscious minds might be accessed by technologically-aided communication using neuroimaging- or neurophysiology-based brain-computer interface (BCI) (Monti et al., 2010, Cruse et al., 2011; Naci et al., 2012; Chatelle et al., 2012). These developments raise the question whether DOC patients (or other behaviorally unresponsive patients) may in fact be able to give their informed consent to treatment or research by means of BCI (Peterson et al., 2013). If patients turn out to be more conscious than presumed they may also retain more autonomous decision-making capacity than presumed and may have to be accorded more rights than they are commonly accorded. In this article, I will reflect on this question and try to give an answer. First, I will briefly present the classic concept of "informed consent" in medical ethics and recapitulate the salient results of the pertinent studies on DOC and awareness. Second, I will show that patients with DOC clearly do not satisfy the traditional requirements of informed consent and that even the assessment of these requirements is problematic in itself. Third, I will argue that the results on DOC patients do not warrant a widening of the concept of informed consent, but should be considered in the context of the concept of assent which can be taken from pediatrics and adapted to the area of neuro-rehabilitation.

Informed consent

Informed consent describes the process of informing about and obtaining permission for an intervention in a person's bodily integrity in the context of health care or research. Without an informed consent, these practices are considered morally unjustified, violating human rights, and even trespassing legal norms in many countries, leading to criminal sanctions (Faden et al., 1986; Miller and Wertheimer, 2009; Maclean, 2013). The duty of informed consent is based on core human rights to self-determination, bodily integrity, and private life. Historically, the necessity of informed consent has been advocated and implemented after the appalling atrocities by the Nazi regime (1933-1945), in particular the human experimentation in concentration camps, and distinctly formulated in the Nuremberg Code (Faden et al., 1986; Nuremberg Code, 1949). In addition, non-consensual experiments on vulnerable persons in other parts of the world (e.g., the Tuskegee experiments) have underscored the importance of informed consent (Lowenstein et al., 2009).

The concept of informed consent consists of three elements: (1) The person has to be adequately and understandably *informed* about the nature, benefits, and risks of the intervention proposed and any alternative options; (2) The person has to possess the cognitive *capacity* to make decisions of the pertinent sort, requiring the abilities to understand adequately given information, to appreciate the decisional situation and its consequences, to reason and weigh the risks and benefits, and to make and communicate a personal decision (American Psychiatric Association, 1998); (3) The person has to actually make the decision *voluntarily*, e.g., without being subjected to coercion, manipulation, or any kind of undue influence (Eyal, 2012). These three preconditions of informed consent – information, decision-making capacity, and voluntariness – have to be assessed in a dialogue with the person. It has to be ensured that the relevant information was offered in a comprehensible way, and if the person did not decline to receive the information, that it was actually understood in the correct way. For the assessment of capacity, a validated structured tool such as the MacArthur Competence Assessment Tool (with separate versions for research or treatment) may be used, but this is commonly only done as an additional support for the judgment of the clinician or researcher (Charland, 2014; Appelbaum and Grisso, 1988; Dunn et al., 2006). Voluntariness, the third requirement, can only be assessed by talking to the person, inquiring his or her motives for participation, and explicitly asking about any possible undue influence.

Informed consent is deliberately conceptualized as a process with relatively demanding requirements to protect the core human rights of the person that are at stake. Thereby, the decision to consent to an intervention in one's bodily integrity should be maximally autonomous. Certain persons, who cannot exercise their autonomy due to a severe cognitive impairment, should be safeguarded from being abused as has frequently occurred in the past. If a person does not satisfy the full criteria for informed consent, the intervention may still be performed (especially as it may be extremely beneficial to that person), but additional and strict safeguards should be installed to prevent abuse: a surrogate with legal decision-making authority should represent the person and give informed consent on his or her behalf, the intervention should have a significant benefit for the person or at least a group with similar characteristics (e.g., with the same illness), and the risks and burdens should be minimal (World Medical Association, 2013). In addition, any efforts should be made to inform the person as much as possible about the proposed intervention and obtain his or her assent in order not to have to perform the intervention against the resistance of the person (compulsorily), which would entail additional risks and the additional infringement of the right to freedom.

Neuroscience on awareness in patients with disorders of consciousness

If patients suffer such severe brain injuries that they remain in a state of DOC (vegetative state or minimally conscious state), they are obviously extremely disabled. By definition, they are unable to initiate a meaningful communication verbally, by mimic or gestures (Bernat, 2006). Their behavioral responsiveness to any stimuli and events in the social world around them is either extremely limited (in minimally conscious state) or absent (in vegetative state). The same is true for spontaneous actions in the traditional sense of the word. They are unable to walk, to eat or drink, to work, to perform leisure activities. Their life critically depends on high-quality around-the-clock nursing care, on artificial nutrition and hydration as well as other forms of life-sustaining medical treatment (Jox, 2011). DOC patients are thus extremely vulnerable to any kind of abuse, from the slightest signs of debasement to the most blatant form of rape (Almodovar, 2002). Moreover, they usually cannot signal any kind of agreement or disagreement with the way they are cared for, the manipulations they are undergoing, or the treatment they are subjected to. Traditionally, patients clinically diagnosed as being in a vegetative state are thought to be completely unaware while patients with the diagnosis of a minimally conscious state are presumed to have phases of rudimentary awareness of parts of their environment or themselves, but no higher forms of consciousness like reflective self-consciousness.

On this background, significant research has been published over the last 10 years. It started with Adrian Owen's case study of a young woman with the clinical diagnosis of a vegetative state, who showed appropriate and distinct cerebral activation patterns in functional magnetic resonance imaging (fMRI) when asked to imagine playing tennis or to navigate through the rooms of her house (Owen et al., 2006). In 2010, a case series of 54 DOC patients was published of whom five (among them four with a vegetative state) displayed the same fMRI reactions, which was interpreted as the willful modulation of brain activity, signaling awareness, understanding of the question, and intentional following of the request (Monti et al., 2010). One patient, clinically being in a vegetative state, even showed appropriate fMRI responses to personal questions when asked to signal yes or no by using different motor imaginations (Monti et al., 2010). A year later, the same group of British and Belgian researchers published that three of 19 patients with a clinical vegetative state were able to follow commands evidenced by appropriate responses in electroencephalography (EEG) when asked to imagine motor behavior (Cruse et al., 2011). A BBC news video published online in 2012 showed a man in a vegetative state who purportedly signaled by motor imageries whether he is in any pain while lying in an fMRI scanner (Walsh, 2012). In a variation of these motor imagery studies, two DOC patients evidently could willfully modulate their responses to personal questions by choosing to attend to auditory stimuli or not (Naci and Owen, 2013).

There are also other methodological approaches to uncover awareness in behaviorally unresponsive and seemingly unconscious patients (Naci et al., 2014;

Hannawi et al., 2015; Demertzi et al., 2015). Generally, these data indicate that the hitherto standard clinical assessment of DOC using the Coma Recovery Scale-Revised may be inaccurate in a small subset of patients, resulting in some aware patients being erroneously held unaware. One of the main questions currently debated is the clinical utility of this neurotechnological assessment of awareness (Peterson et al., 2015). There is also controversy about the benefit that these studies have for the patients and their families (Jox et al., 2012; Graham et al., 2015). More generally, it is debatable whether being minimally aware in a state of impaired responsiveness is better or worse for the patient than being completely unaware (Kahane and Savulescu, 2009; Wilkinson et al., 2009; Levy and Savulescu, 2009; Jox and Kuehlmeyer, 2013). Another major question that arises from these studies concerns the extent to which these technologies allow the assessment of decision-making capacity and the obtainment of informed consent (Lee et al., 2015).

Informed consent via brain-computer interface?

If the use of motor imagery paradigms and fMRI- or EEG-based assessment of patient reactions offer new ways of motor-independent communication, this can be seen as brain-computer-interfaces (BCI) for DOC patients (Shih et al., 2012; Wolpaw and Winter Wolpaw, 2012). BCIs have so far been mainly studied in patients with tetraplegia who are cognitively intact, e.g., patients with spinal cord transections or patients with amyotrophic lateral sclerosis (motor neuron disease) (Kubler and Neumann, 2005; Brunner et al., 2015). The study and use of BCI in patients with severe cerebral cortical injury raises new ethical and legal questions (Chatelle et al., 2012; Naci et al., 2012).

If a robust communication with DOC patients could be established using motor-imagery-based BCIs, this would open up many potential forms of application. First, it may be a window into the subjective state of well-being of that person. Well-being is a cornerstone in any ethical evaluation, and two of the four principles of biomedical ethics (the principles of beneficence and non-maleficence) relate to the patient's well-being. A DOC patient may be asked whether he or she is in pain or any other distress, feeling well or unwell, being happy or sad. For sure, you would always have to critically consider in how far any impairments of speech, cognition, or execution allow the patient to communicate on this level. In many cases, only basic forms of communication may be possible, with short questions referring to concrete actual bodily or emotional states. The more abstract and cognitively demanding the communication will be, the less likely it will be to establish a reliable communication with the patient. So far, studies have only used simple yes-no questions (Fernandez-Espejo and Owen, 2013). While yes-no-signaling can in principle be used to select words and thus form sentences and write elaborate books (analogous to word selection by simple movements as mastered by some locked-in patients) (Dudzinski, 2001), this is unlikely to be successful in DOC patients given their cognitive impairments due to the brain injury.

Second, such BCIs may in principle also be used to probe the patient's cognitive state and decision-making capacity. In a theoretical article, neuroscientists have sketched ways to assess decision-making capacity by employing yes-no questions via BCI (Peterson et al., 2013). This approach, however, suffers from a reductionist concept of autonomy and decision-making capacity. As explained above, decision-making capacity requires understanding, appreciation, reasoning, and communication. In order to assess whether these four elements are present to a sufficient degree, a reliable, nuanced, bidirectional and personal communication is unavoidable (Jox, 2013). As DOC patients cannot read, any information on research or treatment interventions has to be given to them orally. Whether the patient has appropriately understood this information, can only be ascertained by asking him or her to rephrase this information in own words. This may be done by BCI, but it requires a technology that allows the patient to spontaneously formulate own sentences. To assess the process of appreciation and reasoning, it is even more necessary to have the patient formulate sentences, pose questions, give answers.

In traditional capacity assessment, at least if it not only relies on written scales, the non-verbal signs of patients are at least as important as the verbal content. It is a well-known fact that non-verbal signs are more authentic than verbal content and are more trustworthy if the two contradict each other (Malandro et al., 1989). Thus, a physician may tell from a patient's mimic and gestures that he did not understand the diagnostic disclosure at all notwithstanding his verbal assurance that he had understood it.

The third element of informed consent, voluntariness, is similarly difficult to assess via BCI. It can only be assessed if a true conversation is possible and the patient can elaborate about his or her intentions, motives, and objectives. This seems impossible with the current rudimentary BCI technologies for DOC patients. Yet, ensuring voluntariness may be particularly important in this patient group. It is well conceivable that the brain injury, especially if it involves the frontal cortex, led to pathological states of thought, volition, and impulse control that restrict the patient's free will. Another threat to voluntariness comes from the outside: as BCI technology advances and includes bidirectional flow of information between the brain and the computer, it is generally possible to manipulate the brain from the outside via BCI, abrogating the patient's voluntariness. Even without such speculative threats, one can ask the question whether a patient may ever be seen as free (and capable of making voluntary decisions) if he or she is in a state of maximal dependency and the only choice is to consent to the treatment or to die.

In summary, these reflections underscore that it is currently impossible to assess whether the preconditions of informed consent – information, decision-making capacity, and voluntariness – are met in an individual case of a DOC patient (Farisco, 2015). The impossibility of this assessment appears to stem not only from technological shortcomings that may be overcome with time, but also rest upon principled problems. Even if we assume that these three preconditions could be perfectly assessed via BCI, it is improbable that DOC patients will satisfy these criteria for full informed consent. The vast majority of DOC patients do not have small, circumscribed lesions impairing responsiveness like in the case of locked-in patients with brain stem pathologies, but suffered extensive cortex damage from global hypoxic-ischemic injury, severe head trauma, major brain hemorrhage or stroke (Bernat, 2009a). These pathologies are highly likely to affect some of the cognitive and volitional abilities that are necessary for understanding of complex medical or research decisions, decision-making capacity, and voluntary decision making. The presence of awareness and the establishment of communication do not automatically mean that the patient is cognitively capable of exercising autonomy and self-determination. These DOC patients may rather resemble patients with stroke, dementia, or psychiatric disorders than cognitively intact patients locked into their bodies due to some peripheral nerve disease like Guillain Barré syndrome. This fact is evidenced by the severe cognitive and memory disturbances that patients display when they recover and emerge from DOC (Wijdicks, 2006; Voss et al., 2006). The popular illusion that DOC patients are healthy minds in disabled bodies is explained by the fact that the persons were usually cognitively intact before they suddenly slipped into coma due to acute brain injury and further nurtured by fiction movies unrealistically showing the sudden awakening and immediate restoration of the patients' previous life and personality (Wijdicks and Wijdicks, 2006).

How to use brain-computer-interface in DOC patients?

If informed consent cannot be obtained from DOC patients via BCI, one possible consequence could be to change the concept of informed consent. One may advance that novel neurotechnological possibilities require a fundamental alteration of our ethical and legal concepts. This, however, is not self-evident. Why should long-held, well-proven regulations be altered once neuroscience offers new ways of interacting with patients? In fact, lowering the cognitive thresholds for informed consent would jeopardize the protection of autonomy of all persons undergoing medical treatment or research. It would carry the risk that the old paternalism that we have finally overcome would resurface. And for DOC patients specifically it would mean that far-reaching decisions (even on life and death) could be made on false grounds, by mistaking some reactions in neuroimaging paradigms for full expressions of autonomy.

When decisions about research participation or medical treatment have to be made, there is still the need for a suitable surrogate, designated by the patient prior to the brain injury (by donating lasting powers of attorney) or appointed by the court, who will have to make these decisions according to traditional legal and ethical standards. If a valid and applicable advance directive exists, it has to be respected. In all other cases, surrogates will have to make a substituted judgment based on previous expressions of the patient or decide according to the patient's best interests (Bernat, 2009b; Jox et al., 2012). BCI-mediated communication, once it will be established as a reliable means of communication with DOC patients, should be used to inquire about the patient's current well-being. It will be important to explore the need for pain and symptom treatment and to guide the choice and dosage of drugs used to treat these symptoms. In the same vein, it can be used to assess the effect of various forms of non-pharmacological treatment, physiotherapy, occupational therapy, sensory stimulation, diverse forms of rehabilitation and nursing care on the patients' well-being. Several authors have already started to deliberate how we can honor DOC patients' welfare interests by exploring their well-being via BCI (Johnson, 2013; Weijer et al., 2014).

Beyond this use of BCI, patients may also be cautiously approached about any needs and wishes they may have. There should be procedures in place to gauge the level of self-reflection that a patient is able to have and communicate. If needs and wishes are clearly expressed via BCI, the communication may also move towards simple measures of care and treatment. The fact that the patient will not be approached for informed consent can lift off the burden from these conversations. Instead of informed consent, surrogates, health care professionals, and researchers should aim to obtain informed *assent*. This is a concept that has been developed in pediatrics in order to respect the growing capacity of self-determination in children and adolescents and involve them in decision making (Spriggs and Caldwell, 2011). Assent itself is not sufficient to justify research or treatment; it does not replace the need for true informed consent. Striving for assent, however, fosters the child's participation in the research study or treatment course improving the outcomes of these procedures for the well-being of the child. In addition, assent mostly obviates situations of resistance and the need for compulsory measures, preventing psychological traumatization and associated health risks.

In the context of research, assent has an even stronger authority than in the context of treatment. While the refusal to give assent (or plain dissent) may be overridden in exceptional clinical circumstances when treatment is urgently necessary to save the life of the person or protect others, it is not allowed in research. The Declaration of Helsinki states: "When a potential research subject who is deemed incapable of giving informed consent is able to give assent to decisions about participation in research, the physician must seek that assent in addition to the consent of the legally authorized representative. The potential subject's dissent should be respected." This means that research should never be done compulsorily, against the resistance or refusal of the subject.

This concept of informed assent has already been transferred from pediatrics to adult medicine in the case of research or treatment for dementia patients (Black et al., 2010). There is no reason why it should not also be transferred and adapted to the research and treatment of DOC patients. Concerning research, I suggest the same regulation as in pediatric research: assent should be sought via BCI-mediated communication if possible, in addition to informed consent by surrogate decision makers. If the patient clearly dissents via BCI, any research on this patient should not be done. If a research study is ongoing, the Declaration of Helsinki already states that consent should be obtained as soon as possible in the course of cognitive recovery (World Medical Association, 2013). I suggest that even assent should be sought as soon as possible during the recovery of the patient.

In the context of care and treatment of DOC patients, informed assent should also be sought via reliable BCI communication, but dissent should not have the same authority as in research. Assent should be carefully sought and guide medical treatment, but should only be considered as one factor in a multifactorial ethical judgment based on patient well-being and patient autonomy (Jox, 2011).

References

Almodovar, P. 2002. *Talk to her [Online]*. Accessed 20 February 2016 from www.imdb.com/title/tt0287467/.

American Psychiatric Association. 1998. Guidelines for assessing the decision-making capacities of potential research subjects with cognitive impairment. *Am J Psychiatry*, 155, 1649–50.

Appelbaum, P. S., & Grisso, T. 1988. Assessing patients' capacities to consent to treatment. *N Engl J Med*, 319, 1635–8.

Bernat, J. L. 2006. Chronic disorders of consciousness. *Lancet*, 367, 1181–92.

Bernat, J. L. 2009a. Chronic consciousness disorders. *Annu Rev Med*, 60, 381–92.

Bernat, J. L. 2009b. Ethical issues in the treatment of severe brain injury: The impact of new technologies. *Ann N Y Acad Sci*, 1157, 117–30.

Black, B. S., Rabins, P. V., Sugarman, J., & Karlawish, J. H. 2010. Seeking assent and respecting dissent in dementia research. *Am J Geriatr Psychiatry*, 18, 77–85.

Brunner, C., Birbaumer, N., Blankertz, B., Guger, D., Kübler, A., Mattia, D., Del R. Millán, J., Miralles, F., Nijholt, A., Opisso, E., Ramsey, N., Salomon, P., & Müller-Putz, G. R. 2015. BNCI Horizon 2020: Towards a roadmap for the BCI community. *Brain-Computer Interfaces*, 0(0), 1–10.

Charland, L. C. 2014. Decision-making capacity. In Zalta, E. N. (ed.), *The Stanford encyclopedia of philosophy (Fall 2014 edition)*. Accessed 20 February 2016 from http://plato.stanford.edu.

Chatelle, C., Chennu, S., Noirhomme, Q., Cruse, D., Owen, A. M., & Laureys, S. 2012. Brain-computer interfacing in disorders of consciousness. *Brain Inj*, 26, 1510–22.

Cruse, D., Chennu, S., Chatelle, C., Bekinschtein, T. A., Fernandez-Espejo, D., Pickard, J. D., Laureys, S., & Owen, A. M. 2011. Bedside detection of awareness in the vegetative state: A cohort study. *Lancet*, 378, 2088–94.

Demertzi, A., Antonopoulos, G., Heine, L., Voss, H. U., Crone, J. S., De Los Angeles, C., Bahri, M. A., Di Perri, C., Vanhaudenhuyse, A., Charland-Verville, V., Kronbichler, M., Trinka, E., Phillips, C., Gomez, F., Tshibanda, L., Soddu, A., Schiff, N. D., Whitfield-Gabrieli, S., & Laureys, S. 2015. Intrinsic functional connectivity differentiates minimally conscious from unresponsive patients. *Brain*, 138(Pt. 9), 2619–31.

Dudzinski, D. 2001. The diving bell meets the butterfly: Identity lost and re-membered. *Theor Med Bioeth*, 22, 33–46.

Dunn, L. B., Nowrangi, M. A., Palmer, B. W., Jeste, D. V., & Saks, E. R. 2006. Assessing decisional capacity for clinical research or treatment: A review of instruments. *Am J Psychiatry*, 163, 1323–34.

Eyal, N. 2012. Informed consent. In Zalta, E. N. (ed.), *The Stanford encyclopedia of philosophy* (Fall 2012 edition). Accessed 20 February 2016 from http://plato.stanford.edu.

Faden, R. R., Beauchamp, T. L., & King, N. M. P. 1986. *A history and theory of informed consent*. New York: Oxford University Press.

Farisco, M. 2015. "Cerebral Communication" with patients with disorders of consciousness: Clinical feasibility and implications. *AJOB Neurosci,* 6, 44–6.

Fernandez-Espejo, D., & Owen, A. M. 2013. Detecting awareness after severe brain injury. *Nat Rev Neurosci,* 14, 801–9.

Graham, M., Weijer, C., Peterson, A., Naci, L., Cruse, D., Fernandez-Espejo, D., Gonzalez-Lara, L., & Owen, A. M. 2015. Acknowledging awareness: Informing families of individual research results for patients in the vegetative state. *J Med Ethics,* 41, 534–8.

Hannawi, Y., Lindquist, M. A., Caffo, B. S., Sair, H. I., & Stevens, R. D. 2015. Resting brain activity in disorders of consciousness: A systematic review and meta-analysis. *Neurology,* 84, 1272–80.

Johnson, L. S. 2013. Can they suffer? The ethical priority of quality of life research in disorders of consciousness. *Bioethica Forum,* 6, 129–36.

Jox, R. J. 2011. End-of-life decision making concerning patients with disorders of consciousness. *Res Cogitans,* 8(1), 43–61.

Jox, R. J. 2013. Interface cannot replace interlocution: Why the reductionist concept of neuroimaging-based capacity determination fails. *AJOB Neuroscience,* 4, 15–17.

Jox, R. J., Bernat, J. L., Laureys, S., & Racine, E. 2012. Disorders of consciousness: Responding to requests for novel diagnostic and therapeutic interventions. *Lancet Neurol,* 11, 732–8.

Jox, R. J., & Kuehlmeyer, K. 2013. Introduction: Reconsidering Disorders of Consciousness in Light of Neuroscientific Evidence. *Neuroethics,* 6, 1–3.

Kahane, G., & Savulescu, J. 2009. Brain damage and the moral significance of consciousness. *J Med Philos,* 34, 6–26.

Kubler, A., & Neumann, N. 2005. Brain-computer interfaces – the key for the conscious brain locked into a paralyzed body. *Prog Brain Res,* 150, 513–25.

Lee, G., Byram, A. C., Owen, A. M., Ribary, U., Stoessl, A. J., Townson, A., Stables, C., & Illes, J. 2015. Canadian perspectives on the clinical actionability of neuroimaging in disorders of consciousness. *Can J Neurol Sci,* 42, 96–105.

Levy, N., & Savulescu, J. 2009. Moral significance of phenomenal consciousness. *Prog Brain Res,* 177, 361–70.

Lowenstein, P. R., Lowenstein, E. D., & Castro, M. G. 2009. Challenges in the evaluation, consent, ethics, and history of early clinical trials – Implications of the Tuskegee "trial" for safer and more ethical clinical trials. *Curr Opin Mol Ther,* 11, 481–4.

Maclean, A. 2013. *Autonomy, informed consent and medical law. A relational challenge.* Cambridge, UK: Cambridge University Press.

Malandro, L. A., Barker, L. L., Gaut, D. A., & Barker, D. A. 1989. *Nonverbal communication.* New York: Random House.

Miller, F. G., & Wertheimer, A. 2009. *The ethics of consent: Theory and practice.* New York: Oxford University Press.

Monti, M. M., Vanhaudenhuyse, A., Coleman, M. R., Boly, M., Pickard, J. D., Tshibanda, L., Owen, A. M., & Laureys, S. 2010. Willful modulation of brain activity in disorders of consciousness. *N Engl J Med,* 362, 579–89.

Naci, L., Cusack, R., Anello, M., & Owen, A. M. 2014. A common neural code for similar conscious experiences in different individuals. *Proc Natl Acad Sci U S A,* 111, 14277–82.

Naci, L., Monti, M. M., Cruse, D., Kubler, A., Sorger, B., Goebel, R., Kotchoubey, B., & Owen, A. M. 2012. Brain-computer interfaces for communication with nonresponsive patients. *Ann Neurol,* 72, 312–23.

Naci, L., & Owen, A. M. 2013. Making every word count for nonresponsive patients. *JAMA Neurol,* 70, 1235–41.

Nuremberg Code. 1949. *Trials of war criminals before the Nuremberg Military Tribunals under Control Council Law No. 10.*, Washington, D.C: Government Printing Office.

Owen, A. M., Coleman, M. R., Boly, M., Davis, M. H., Laureys, S., & Pickard, J. D. 2006. Detecting awareness in the vegetative state. *Science*, 313, 1402.

Peterson, A., Cruse, D., Naci, L., Weijer, C., & Owen, A. M. 2015. Risk, diagnostic error, and the clinical science of consciousness. *Neuroimage Clin*, 7, 588–97.

Peterson, A., Naci, L., Weijer, C., Cruse, D., Fernandez-Espejo, D., Graham, M., & Owen, A. M. 2013. Assessing decision-making capacity in the behaviorally nonresponsive patient with residual covert awareness. *Am J Bioeth Neuroscience*, 4, 3–14.

Shih, J. J., Krusienski, D. J., & Wolpaw, J. R. 2012. Brain-computer interfaces in medicine. *Mayo Clin Proc*, 87, 268–79.

Spriggs, M., & Caldwell, P. H. 2011. The ethics of paediatric research. *J Paediatr Child Health*, 47, 664–7.

Voss, H. U., Uluc, A. M., Dyke, J. P., Watts, R., Kobylarz, E. J., Mccandliss, B. D., Heier, L. A., Beattie, B. J., Hamacher, K. A., Vallabhajosula, S., Goldsmith, S. J., Ballon, D., Giacino, J. T., & Schiff, N. D. 2006. Possible axonal regrowth in late recovery from the minimally conscious state. *J Clin Invest*, 116, 2005–11.

Walsh, F. 2012. Vegetative patient Scott Routley says "I'm not in pain." *BBC News*. Accessed 20 February 2016 from www.bbc.com/news/health-20268044.

Weijer, C., Peterson, A., Webster, F., Graham, M., Cruse, D., Fernandez-Espejo, D., Gofton, T., Gonzalez-Lara, L. E., Lazosky, A., Naci, L., Norton, L., Speechley, K., Young, B., & Owen, A. M. 2014. Ethics of neuroimaging after serious brain injury. *BMC Med Ethics*, 15, 41.

Wijdicks, E. F. 2006. Minimally conscious state vs. persistent vegetative state: The case of Terry (Wallis) vs. the case of Terri (Schiavo). *Mayo Clin Proc*, 81, 1155–8.

Wijdicks, E. F., & Wijdicks, C. A. 2006. The portrayal of coma in contemporary motion pictures. *Neurology*, 66, 1300–3.

Wilkinson, D. J., Kahane, G., Horne, M., & Savulescu, J. 2009. Functional neuroimaging and withdrawal of life-sustaining treatment from vegetative patients. *J Med Ethics*, 35, 508–11.

Wolpaw, J., & Winter Wolpaw, E. 2012. *Brain-computer interfaces: Principles and practice.* Oxford and New York: Oxford University Press.

World Medical Association. 2013. *Declaration of Helsinki – Ethical Principles for Medical Research Involving Human Subjects* [Online]. Accessed 20 February 2016 from www.wma.net/en/30publications/10policies/b3/.

10 Brain-imaging and privacy concerns

Arleen Salles

Introduction

Rapid advances in functional brain-imaging technology during the last decade have enriched our knowledge of the living brain. Electroencephalography (EEG) allows the detection of the electrical activity of millions of neurons in different cortical regions, positron emission tomography (PET) allows the identification of metabolic activity to particular areas of the brain by injecting the subject with radioactively labeled elements, and functional Magnetic Resonance Imaging (fMRI) allows the mapping of neural function by measuring correlated level of blood oxygenation in the brain (Aguirre, 2014). These and other techniques make both the measurement of different structural features in the brain and a view of the brain in action possible: they have been used for the identification and analysis of disease processes, pre-surgical mapping, and the detection and assessment of residual consciousness in patients who have traumatic brain injuries (Racine and Illes, 2007; Evers and Sigman, 2013). They could also be used to communicate with minimally conscious patients (Bendtsen, 2013) and even as a tool for decision making of patients suffering from disorders of consciousness (Peterson et al., 2013). The wide availability and non-invasive character of fMRI (as compared to other imaging techniques such as PET) together with its considerable spatial resolution, make it particularly appealing as an instrument to use in a variety of non-clinical contexts as well. This technique measures localized changes in the brain as an individual is performing particular tasks and is typically used to gain understanding about social cognition, affective processes, and brain activity in general. Recently, neuroscientists have used fMRI techniques when trying to identify brain regions associated with deception (Farah et al., 2014; Stoller and Wolpe, 2007; Langleben et al., 2006; Wolpe et al., 2005), violent and criminal behaviors (Bufkin and Luttrell, 2005), and empathy (Decety et al., 2013), and to explain moral decision-making and judgment (Greene, 2014), among others.

In the research and clinical contexts, imaging procedures in general raise the familiar ethical issues of informed consent, how to deal with incidental findings, confidentiality, and how to manage potentially sensitive information. This is true of brain imaging as well (Kulynych, 2002; Greely, 2004; Alpert, 2005; Tovino, 2005; Farah, 2012; Edwards, 2012). However, some non-clinical uses of fMRI based brain imaging raise additional and more unique moral challenges.

This technique combined with complex statistical analysis allows a more detailed decoding of people's mental states and thus presents a novel threat to privacy (Haynes, 2011).

Concerns about privacy and how to respect it are not specific to neuroscience. Respect for privacy in both the research and the clinical context is typically recognized as a basic bioethical requirement. Indeed, citizens in democratic societies typically expect to have a certain degree of privacy (that they generally articulate in terms of the right to keep information to themselves and have control over who will have access to it). That this is a reasonable expectation (both descriptively and normatively) is shown by the existence of a number of legal safeguards (such as rules of protection of data and formal recognition of privacy rights) that are applied under normal circumstances.

However, there is the fear that some brain-imaging based neuroscientific studies will pose a novel threat, in so far as they might provide insights into other people's thoughts, perceptions, and emotions in the absence of outwardly observable behavior or speech. In the process, not only would they undermine the self-conception that humans have as being uniquely positioned to know their own mental content: their findings could be used for a variety of disturbing ends.

Already some fMRI studies have been able to provide psychological information about people, including information about specific intentions, visual responses, unconscious biases, and intentional deception (Wolpe et al., 2005; Haynes and Rees, 2006; Farah et al., 2008; Haynes, 2011, 2012). By identifying correlations between brain and mental activity, brain-imaging technology might further restrict the private sphere and do so incrementally as time goes by. What would this mean from an ethical standpoint? Would this be a problem? And how to address it?

In this chapter, I examine these issues. But first a caveat. I do not focus on the matter of privacy inside the clinical and research settings. This means that I do not directly address questions widely discussed in those contexts, for example, technical legal measures and controls designed to protect privacy. I concentrate on the issue of mental privacy insofar as functional neuro-imaging can reveal information about people's mental states and psychological traits but I do not specifically address the issue of mental privacy in recent research on minimally conscious patients. I believe, however, that unless one holds a simplistic view according to which an individual is either permanently unconscious or normally conscious, many of the points made here may be quite relevant to the discussion of mental privacy of minimally conscious patients.

In the following, I highlight some efforts to approach the issue of functional neuro-imaging and its possible threat to privacy in the neuroethics literature. Two main approaches or strategies are usually used in the discussion: the first strategy consists in a description and discussion of what neuro-imaging can and cannot do with a focus on the technical and methodological problems that bedevil the technology. The second strategy focuses on the metaphysical assumptions about the mind underlying concerns on the subject of neuro-imaging and mental privacy. Sometimes these two strategies are used jointly. There is a third strategy, less common in the neuroethics literature, that brackets technical, methodological, and metaphysical issues to put the focus on the discussion of normative questions. The questions raised are: why would neuroimaging's impinging on privacy be problematic? What is valuable about mental privacy? Would it be morally undesirable to have less of it? My main aim is to outline the first two strategies clarifying their implications for the privacy debate, and then focus more on the third. I end by proposing to expand the normative discussion to incorporate some of the issues raised by a recent account of privacy as contextual integrity.

First strategy: focusing on technical and methodological limitations

The prospect of reading people's thoughts causes great alarm in the public (Lever, 2012; Shen, 2013; Miller, 2014; Golgowski, 2014). To assuage fears and stop unrealistic thinking about the likelihood of achieving a general "brain reading device," several authors have presented and explained the technical and methodological issues that plague the technology (Wolpe et al., 2005; Buller, 2005; Haynes and Rees, 2006; Racine and Illes, 2007; Racine et al., 2010; Haynes, 2011, 2012; Evers and Sigman, 2013). They note that it is extremely doubtful that universal brain reading (that is, decoding arbitrary mental states belonging to arbitrary people) will be available in the near future, if ever feasible and usable. (Racine and Illes, 2007; Racine et al., 2010; Haynes, 2011, 2012, among others) The possibility of accurate reconstruction of mental states from measurements of brain activity and the usability of such technique depends on solving three main challenges not easy to overcome: those presented by the experimental designs and the technology itself, those derived from the organization of human brains, and those presented by the need to interpret the scientific data.

There is scientific agreement that FMRI capabilities to decode mental states are quite limited by a number of factors (Logothetis, 2008). While its spatial resolution is unequalled when compared to other imaging techniques, its temporal resolution is insufficient to allow the differentiation of subtly different patterns of brain activity and the correlating mental states (Haynes and Rees, 2006; Gallant Lab, 2011; Haynes, 2011, 2012; Aguirre, 2014). Furthermore, it is necessary to increase its sensitivity: fMRI signals are contaminated by strong noise and background physiological fluctuations and this affects the extraction of subtle functional information. Other limitations are related to the fact that the accuracies reported in the fMRI studies are based on artificial lab tasks different from real life situations and settings, and require the participation of cooperative subjects (Haynes and Rees, 2006; Haynes, 2011; Farah et al., 2008; Aguirre, 2014).

Yet, even if the above-mentioned limitations were overcome, there are additional problems related to what is being measured. Most human decoding studies are carried out under highly simplified conditions making generalization across time, subjects, and situations, and generalization across different instances of the same mental content questionable (Haynes and Rees, 2006; Farah et al., 2008; Racine et al., 2010; Aguirre, 2014). Moreover, although some research shows that patterns of brain activity for some thoughts are similar across people (for example, those coding whether somebody is lying), different brains code information in subtly different ways. Brain variability responds not only to biology but also to psychological and socio-cultural factors (Chiao and Immordino-Yang, 2013; Evers and Sigman, 2013; Haynes, 2012). The neural difference exhibited means that in order to read a person's brain, it would be necessary to know how that person's brain codes the relevant mental states (Haynes, 2011, 2012).

Finally, most commentators stress the challenge presented by interpretation of the data (Racine et al., 2005; Illes and Racine, 2005; Haynes and Rees, 2006; Farah et al., 2008; Racine et al., 2010). Decoding thoughts requires inferring mental meaning from surrogate physiological signals (specifically changes in blood oxygen level) that are poorly understood (VanMeter, 2010). BOLD changes depend not only on neuronal activity but also on other variables such as age or disease (Schleim and Roisier, 2009). Furthermore, the decoding is importantly based on what is known as "reverse inference," that is, the assumption that because specific brain regions are particularly active during a specific mental state, activation of such brain regions means presence of that mental state. Although not invalid, the usefulness of such inference is limited (Poldrack, 2006; Farah, 2014). Other concerns revolve around the problematic equation of correlation and causation.

It is true that the discussion of the limits of any technology is very important when reflecting on its social and ethical implications (Roskies, 2007; Fins, 2011; Parens and Johnston, 2014). This becomes evident when one focuses on the multiple current and sometimes premature applications of fMRI technology. It is hard to disagree with those who warn us regarding its limitations. However, a recognition of such limitations should entail neither the dismissal of the technology (Farah and Hook, 2013; Farah, 2014) nor the belief that some of the main normative issues raised by it are solved. First, as it has been pointed out, for each of the technology's practical obstacles there is new research in the process of overcoming it (Aguirre, 2014; Farah, 2014). And second, talk about fMRI's shortcomings does not show that the technology is immune to privacy concerns. To be fair, in general, those authors who point out the limitations of the technology still underline that neuro-imaging will have some impact on privacy and many are clearly aware of future regulatory challenges as the technology advances. However, this strategy remains fundamentally descriptive and because its main objective is to highlight scientific and methodological limitations, concerns about privacy often appear as an afterthought. For the same reason, this strategy either takes the value of privacy for granted or just leaves the issue of its value unexamined.

Second strategy: metaphysical assumptions

The existence of brain decoders is not a specific neuroscientific objective at this point and we have seen that there are limits to what the technology can accomplish at present. However, while it might be comforting to think of current limitations, it is not unreasonable to think that the technology will move beyond its current state, overcoming some of its technical and methodological shortcomings. So let's assume that brain reading devices were possible … what would this mean for mental privacy? A handful of commentators believe that this question cannot be answered without first understanding the difference between reading brain activity and reading mental activity.

Indeed, some authors criticize what they believe to be the general tendency to lose sight of a metaphysical assumption underlying the view that brain reading will threaten mental privacy: that the mental is reducible to the physical; that subjectivity can be reduced to brain activity (Racine et al., 2005; Buller, 2005; Racine et al., 2010; Gilead, 2015).[1] This assumption, it is argued, is highly problematic.

Without rejecting the idea that there is a correlation between psychological and neurobiological states, a few authors note that at least some subjective mental contents still remain impenetrable to others.[2] To illustrate, Kathinka Evers and Mariano Sigman state that "subjectivity introduces an unknowable realm in the world of every individual, as impressions of another – and maybe of oneself – always pass through a filter of subjective interpretations" (2013, p. 891). It is a logical point: there is a domain of subjective experience that cannot be fully transferred to others, nor can it be shared. To that extent, some mental privacy will always exist. Evers, however, does not deny the possibility of some objective measurement of the subjective experience of individuals. The author draws on the work of neuroscientists Joseph Ledoux, Jean Pierre Changeux, and Gerald Edelman to present a dynamic view of the brain that she calls "informed materialism." According to this view, adequate understanding of our subjective experience must take *both* self-reflective information and data gathered from physiological observations and physical measurement into account. Of course, such measurement will always be limited by the "complexity and plasticity of subjective experience, the inevitable filter of intersubjective interpretation" and the issue of what is a legitimate language to use in interpreting results (Evers, 2009, 2010, p. 55). That is, even if imaging techniques could penetrate our thoughts in the sense of delivering general knowledge about mental states, they would not be able to deliver knowledge of the specific meaning of such mental state for the person precisely because such knowledge is subjective and can only be acquired through first person experiences. Functional brain imaging might illuminate the neural underpinnings of thoughts but is unlikely to provide the experience itself. This means that there is always going to be a private mental sphere, even if more and more limited as the technology is developed further. But Evers is not in principle averse to the possibility of brain decoders eventually having access to mental content, even inner content to which the self may have no introspective access (Evers and Sigman, 2013, p. 895).

A different position regarding the possibility that brain reading could ever deliver any useful information on mental content can be found in a recent paper by philosopher Amihud Gilead. Like others, Gilead holds that the kind of mental activity truly constitutive of the subject is essentially subjective. This means that it cannot be shared or transferred to others. If subjectivity were shareable, he adds, it would be objective, and therefore it would lose meaning, becoming ultimately insignificant (Gilead, 2015). In contrast with Evers, however, he sees the subjective, the inter-subjective, and the objective as different types of reality, irreducible to one another even if necessarily connected to each other. For Gilead, it is not a question of points of view, rather of different *natures* and their attributes: the subjective, irreducible and impenetrable is essential to and only accessible to the self,[3] while the objective and intersubjective are publicly penetrable. In short, even if the brain and the mind are inseparable, the mind is available only to itself. Thus brain reading, understood as a means to access the brain, does not imply mental reading which would entail epistemic accessibility to the minds of others.

Gilead's view is intended to have specific implications for the issue of mental privacy. He believes that it should mitigate some of people's fears regarding the possibility that brain reading will actually encroach on mental privacy. Revealing brain activity would not entail revealing fundamentally subjective information: there is no epistemic access to a person's experiences and mental states because those experiences and states are different from physical states occurring at the same time. Gilead explicitly states that the concern about mental privacy is thus unjustified. An implication of his view appears to be that only encroachments of mental privacy deserve our attention and concern, thus suggesting that a breach of brain privacy is less potentially troubling than a breach of mental privacy (or rather, that it would only be troubling to those who hold a strict neuroessentialist view according to which objective knowledge about neuronal activities in the brain is enough to fully explain the self).[4] Now, while it may be true that different views about the mind-brain relationship might play a role in how strong people's concerns about neuro-imaging breaching mental privacy are, mental privacy concerns may arise even if neuro-imaging does not fully reveal one's subjective thoughts or feelings. Sarah Stoller and Paul Root Wolpe note that "if we view our minds as our 'selves' and our brains as enabling our minds, then technologies capable of uncovering cognitive information from the brain threaten to violate our sense of privacy in a new and profound way" (Stoller and Wolpe, 2007, p. 372). But note that this does not have to presuppose a reductionistic view according to which brains states are mental states and mental content can be easily read by looking at fMRI results (Kahane, 2008). Insofar as neuroscience has identified many empirical correlations between brain activity and mental life, it makes sense to think that we might be able to know enough of relevance about the mental lives of human beings (even if we do not have a completely accurate reading of their minds). In short, we do not need to assume that brain reading will produce exact reports of psychological states to believe that it might be able to tell us something about the mind and therefore that mental privacy is not totally guarded against it.

Third strategy: unveiling normative considerations

A combination of the first two strategies is evident in many neuroethics articles on neuroimaging that relate to mental privacy. The main gist is both that the technology is limited and that regardless of the future sophistication of the current tools there will always be a certain degree of privacy (more or less limited). Of course, this does not negate the importance of regulation, but even anticipating neuro-technological advances, ultimately "putting an MRI scanner in the police station will not trample on our mental freedoms because the complexity of the mind-brain relationship will prevent the government from using brain data to reliably read individual minds" (Shen, 2013).

Probably so, but do empirical and metaphysical considerations exhaust the discussion on brain imaging and privacy? It may very well be that neuro-imaging technology will never be able to deliver wholly reliable and accurate information on people's thoughts. However, the normative issue is still open. If brain-imaging technology advanced to read our thoughts, or at least some of them, would that be morally troubling? And if so, why? To answer this question, we must move beyond empirical data and metaphysical views on the mind brain relationship. This is what a third strategy does: it concentrates on issues such as whether mental privacy gains and losses are morally desirable, how to understand privacy, whether there is a moral right to privacy and what this means, and how to assess techniques or practices that limit privacy.

In a recent paper, philosopher Sarah Richmond concentrates on the issue of whether mental privacy losses might be valuable. She asks her readers to imagine a "transparency scenario" where people would be allowed to wear mind-reading glasses to read the thoughts of whoever they look at. In this scenario, everybody is equally exposed to such mind reading and thus equally vulnerable. One could only protect one's mental content by (a) requesting others not to wear mind reading glasses, (b) hiding, or (c) producing "screen" thoughts. It is clear that in this scenario people would have access to information typically unavailable (Richmond, 2012).

Some of us might think that such scenario is undesirable both from a prudential and moral standpoint. Richmond, however, wants to show the opposite: the lack of mental privacy that could ensue from advances in neuroimaging might not be as objectionable as initially thought.

Richmond distinguishes among three types of fairly sensitive thoughts that people have a tendency to want to keep to themselves: preferences and likes that they keep in order to protect their image; lies and forms of deception used for securing certain personal advantages; and finally, thoughts that, if made public, would potentially harm others. Exposing many of these, Richmond believes, may actually be more beneficial than harmful. First, people often want to keep their preferences and likes private because such preferences are unusual or at odds with socially shared norms (imagine, for example, those who keep their sexual orientation private). However, Richmond argues that such protective self-censorship in itself might be deeply troubling for a number of reasons, among them that insofar as those preferences are kept private, they cannot be used to examine and even possibly revise problematic socially shared norms. In that sense, thought transparency might foster needed reflection on societal norms and principles and in the process increase tolerance towards the views of others.

Of course, one might want to keep preferences to oneself not because they contradict socially shared norms but rather because they are incompatible with one's image or the image that one wants to convey (for example, the erudite Oxford professor who prefers to keep her preference for reality TV private). Again, Richmond argues that this "keeping appearances" is equally problematic because it is a sign of self-contempt. Exposure might be a solution insofar as it might foster a healthier self-understanding and definition.

According to Richmond, the advantages of the transparency scenario become more evident when it comes to exposing wrongful lies. Mental transparency is quite likely to reduce crime: it will make it more difficult for people to take advantage of others by not providing accurate information. The author is not worried that thought exposure would entail a violation of potential liars' autonomy. She claims that the transparency scenario would not force people to tell the truth: it would just make lying more difficult.

Richmond is aware that the benefits of mental transparency are not clear when it comes to the third kind of thought that people prefer to keep private: white lies and keeping one's feelings secret to spare pain or suffering to others. Relationships such as romantic love, she notes, need to be further examined to see whether they might be improved by mental transparency. She also recognizes that thought transparency would affect personal and interpersonal expectations which, at least initially, could be problematic. But she is optimistic about the possibility of overcoming such potential problems. Ultimately, she wants us to consider that "the alarm with which many people respond to the prospect of mind reading is unnecessary" (Richmond, 2012).

Richmond's transparency scenario serves an important goal: it makes us look at some of our private preferences in a different light, not necessarily as autonomy enhancing but rather as constraints that we set on ourselves just to fit in. In some of the cases she mentions more thought exposure might be, all things considered, better and might promote a different, morally healthier society. A shortcoming of her account, however, is that she does not seriously consider the possibility that the alarm with which people respond to the possibility of mind reading might also be related to the idea that privacy is valuable in itself, regardless of its possible consequences. An additional problem is that the scenario Richmond presents risks cheating its way around the real concerns that people feel when considering the prospect of mind reading. It does so by stipulating the existence of a "benign government" that makes the technology available to all, that is, where everybody is equal with regards to brain reading capabilities and vulnerabilities. Richmond's transparency scenario also suggests some type of consent since she stipulates that there are a few means that one could select to keep mental content hidden from others (thus suggesting that people would choose it).[5] However, the problem is that even if we ask people to imagine a world where mind reading glasses are scientifically possible and available (that is, even if we ask them not to stay grounded in what is scientifically likely to develop) it is unlikely that the proposed scenario will appear realistic enough. In more realistic hypothetical scenarios, the mind reading technology would be in the hands of a few government officials, for example, who would use the flows of information for surveillance and other sinister purposes such as social control. Or the technology would be in the hands of private corporations that would monitor people's preferences and use such knowledge for their own advantage. Even further, in a more realistic hypothetical scenario, regular people would be powerless to stop those who do the mind reading, either because legal safeguards are not in place, or because even if they are, those in power find a way around them. It is unlikely that people would not imagine the many possible ways in which such technology could be misused.

In short, considering existing power inequalities and what history and science tell us about human beings generally self-interested behavior, the proposed transparency scenario does not seem a completely adequate point of reference to convince people of the advantages of mind reading devices and of the corresponding loss of privacy. Even if thought provoking, the scenario is unlikely to make sense of the fears that people have, help mitigate such fears or allow us to determine how to understand privacy losses if mind reading technology develops.[6]

A key to understanding what could be at stake if mind reading were possible, and people's reactions to potential privacy losses might require revisiting the normative notion of privacy (that is, the value we place on privacy). However, the problem is that this is an essentially contested issue: there is no philosophical agreement on how to understand it and on where to locate its value. Some theoretical work sees privacy valuable as a means to attain a desirable end such as well-being and autonomy, the development and cultivation of close relationships, and true social and political citizenship.[7] Others see privacy as intrinsically desirable, an essential component of human dignity itself. In general, there is a tendency to relate privacy to self-determination and control over who has access (informational and physical) to the self. Can these views help us understand what is at stake? Considering technological advances, does privacy need to be reconceptualized?

Recently, Helen Nissenbaum (2004, 2010) has provided a positive answer to this last question. She argues that existing scholarship on privacy fails to provide adequate responses to many of the privacy challenges raised by contemporary technology-based systems and practices. Nissenbaum does not focus specifically on brain scanning and mental privacy: rather she provides a conception of privacy intended to solve some of the concerns raised by the new technologies. Her account could be used, I think, to inform the discussion on privacy concerns raised by neuro-imaging.

Nissenbaum is convinced that we need an account of privacy that will capture the reason why people have a negative reaction to a diminishment of privacy and provide a justificatory platform to explain and assess the privacy concerns raised by emerging technologies. For Nissenbaum, people's main concern with the new technologies and their impact on privacy does not revolve simply around issues of access but rather and fundamentally around issues of appropriateness. Privacy is about appropriate flows of information, Nissenbaum argues, and a right to privacy is not a right to control of information but a right to appropriate flow of information, that is, a right to have one's expectations regarding how information must flow met. This is what she calls "contextual integrity" which is "preserved when informational norms are respected and violated when informational norms are breached" (Nissenbaum, 2010, p. 140).

The key parameters of those informational norms are actors (that is, who sends the information, who receives such information, and who is the information about); attributes (what is the nature of the information in question); and transmission principles such as, for example, confidentiality, reciprocity, and entitlement that express terms and conditions under which transfers of information should or should not occur (Nissenbaum, 2010, p. 145). According to Nissenbaum, answering questions about violations of privacy means asking questions about whether a particular practice violated context-relative informational norms. The answer calls for a determination of the relevant context, an explanation of the relevant informational norms, and an identification of disruptive flows which must be evaluated against such context relative informational norms. "If the new practice generates changes in actors, attributes, or transmission principles, the practice is flagged as violating entrenched informational norms and constitutes a prima facie violation of contextual integrity," Nissenbaum adds (p. 150). People's disquiet when thinking about some of the new technologies is thus explained by the fact that such technology may be at odds with expectations people have regarding how information will be transmitted in their social context. Many new technologies contravene informational norms and thus affect the whole social fabric (p. 148). Contextual integrity, then, explains the sources of the concern and clarifies when a practice has actually violated privacy.

A more interesting and challenging question, however, is whether such violation of privacy is justified. After all, some disruptions of the flow of information might actually be good. Nissenbaum is open to this possibility. She states that her contextual integrity framework is intended to provide guidance to evaluate practices that contravene informational norms. For this she calls for a consideration of how such practices impact moral and political values and how they interfere with the values, goals, and ends of a particular context (p. 182). It is on the basis of those further evaluations that the framework can recommend acceptance or rejection of the practice in question (p. 183).

As noted before, Nissenbaum does not discuss mental privacy and brain reading in her work. However, it would be interesting to see whether her contextual integrity framework can enrich the thinking on the normative issues involved in this area. It is not my intention to provide a full-fledged analysis of this possibility here but rather to outline possible lines of inquiry. It seems that if we were to use contextual integrity, we would need to address a number of questions such as: if brain reading could actually deliver mental content, what would the violation of privacy exactly consist in? Would it be related to a change in the actors, that is those who give, receive, and are subjects of the information? Would such breach of privacy be connected to the kind of information involved: that is, could it be said that we would be dealing with a new type of information (no longer *about* the self "but rather of the actual self" [Stoller and Wolpe, 2007, p. 372]) and we are uncertain as to what this means? Or would such privacy violation be related to the alteration of transmission principles? In short, it seems that a fundamental question to answer would be: what is the relevant context, and what would be the context-relative informational norms that could be violated by neuro-imaging?

On Nissenbaum's approach, finding an answer to this would already allow us to understand the reason for the concern and even panic that many people feel when they think of the possibility of brain reading. Next, we would have to determine the legitimacy of such privacy-encroaching practice. Who is harmed by it and how? Specifically, what kinds of threats does it present to the ends and objectives of the specific context? (Nissenbaum, 2010, p. 182). On the basis of the discussion, we would be able to think more clearly about the meaning and seriousness of the privacy concerns raised by neuro-imaging.

Conclusion

The issue of the privacy implications of brain imaging will continue to be problematic and produce fears in many. Here, I presented three strategies used to address the main concerns. The first focuses on scientific and methodological considerations and tends to conclude that even if some worry about brain-imaging diminishing privacy is justified and merits attention, many of the most troubling scenarios in the popular imagination are science fiction.

A second strategy revolves around the discussion of the metaphysical assumptions underlying the belief that brain reading will entail lack of mental privacy. Several authors suggest that even if brain imaging allowed full access to the brain, this would not mean that it would allow full access to the mind. Thus, we should not be as worried about the possibility that brain imaging will substantially infringe on our mental privacy.

Useful as they are, these two strategies do not wholly settle people's concerns regarding privacy. They leave out several normative issues, including those about the meaning of privacy, and the values behind people's reactions when confronted with the possibility of a future in which mind reading is possible. I presented the views of two authors who directly or indirectly try to unveil some of those normative concerns.

I believe that future efforts should be directed towards a deeper discussion of the normative issues. My suggestion should not be interpreted as defending the view that methodological and metaphysical considerations are unnecessary in the discussion. In fact, I consider them key when reflecting on these issues. My point is rather that we need a more complex analysis of normative issues in order to be able to reach conclusions regarding the seriousness of brain imaging's threat to privacy.

Notes

1 The issue of the relation between neural and mental states is giving new currency to an old philosophical debate about the mind and the body. This issue requires more analysis than is possible here.
2 It has also been suggested that even if the brain's biology could be perfectly understood and mental states made transparent, we would still need more to understand person's thoughts. "An unedited transcript of someone mental processes is likely to present deep problems of intelligibility that cannot be overcome without the

thinker's assistance [...] our experience of ourselves suggest that the bulk of these transcripts will be obscure, often misleading and numbingly boring" (Richmond, 2012, p. 191). The main point is that those thoughts and preferences make sense in relation to something else, and it is unlikely that the technology will give us access to that.

3 This is certainly a controversial view as well: it is not self evident that the subjective is always accessible to the self. See Evers (2010).

4 I thank Kathinka Evers and Michele Farisco for raising this issue.

5 The issue of whether consent is enough to protect privacy requires analysis that I do not provide here. For a discussion of this question, see Edwards (2012).

6 I am not saying that people's emotional reactions constitute moral arguments but rather that they can be a useful starting point when trying to determine why a loss of mental privacy might be morally problematic.

7 For classic accounts, see for example Fried, 1970, and Rachels, 1975.

References

Aguirre, G. 2014. Functional neuroimaging: Technical, logical, and social perspectives. *Hastings Center Report*, 45(2), S8–S18.

Alpert, S. 2005. Brain privacy: How can we protect it? *AJOB*, 7(9), 70–3.

Bendtsen, K. 2013. Communicating with the minimally conscious: Ethical implications in end of life care. *AJOB Neuroscience*, 4(1), 46–51.

Bufkin, J., & Luttrell, V. 2005. Neuroimaging studies of aggressive and violent behavior. *Trauma, Violence and Abuse*, 6(2), 176–91.

Buller, T. 2005. Can we scan for truth in a society of liars? *The American Journal of Bioethics*, 5(2), 58–60.

Chiao, J. Y., & Immordino-Yang, M. H. 2013. Modularity and the cultural mind: Contributions of cultural neuroscience to cognitive theory. *Perspectives on Psychological Science*, 8, 56–61.

Decety, J., Chen, C., Harenski, C., & Kiehl, K. A. 2013. An fMRI study of affective perspective taking in individuals with psychopathy: Imagining another in pain does not evoke empathy. *Frontiers in Human Neuroscience*, 7, 489.

Edwards, S. 2012. Protecting privacy interests in brain images: The limits of consent. In Richmond, S., Rees, G., Edwards, S. (eds.), *I know what you are thinking: Brain imaging and mental privacy*. Oxford: Oxford University Press.

Evers, K. 2009. *Neuroéthique. Quand la matière s'éveille* [*Neuroethics: When matter awakens*]. Paris: Éditions Odile Jacob. Accessed 12 February 2016 from www.crb.uu.se/downloads/Evers-Neuroethique.pdf.

Evers, K. 2010. *Neuroetica: Cuando la material se despierta* [*Neuroethics: When matter awakens*]. Buenos Aires: Katz Editores.

Evers, K., & Sigman, M. 2013. Possibilities and limits of mind reading: A neurophilosophical perspective. *Consciousness and Cognition*, 22, 887–97.

Farah, M. 2012. Neuroethics: The ethical, legal, and societal impact of neuroscience. *Annual Review of Psychology*, 63, 571–91.

Farah, M. 2014. Brain Images, babies, and bathwater: Critiquing critiques of functional neuroimaging. *Hastings Center Report*, 45(2), S19–30.

Farah, M., & Hook, C. 2013. The seductive allure of "Seductive Allure." *Perspectives on Psychological Science*, 8(1), 88–90.

Farah, M., Hutchinson, J. B., Phelps, E. A., & Wagner, A. D. 2014. Functional fMRI based lie detection: Scientific and societal challenges. *Nature Reviews Neuroscience*, 15(2), 123–31.

Farah, M., Smith, M. E., Gawuga, C., Lindsell, D., & Foster, D. 2008. Brain imaging and brain privacy: A realistic concern? *Journal of Cognitive Neuroscience*, 21(1), 119–27.

Fins, J. 2011. Neuroethics and the lure of technology. In Illes, J., & Sahakian, B. (eds.), *The Oxford handbook of neuroethics*, pp. 895–907. New York: Oxford University Press.

Fried, C. 1968. Privacy: A moral analysis. *Yale Law Journal*, 77(1), 475–92.

Gallant Lab. 2011. Reconstructing visual experiences from brain activity evoked by natural movies. Accessed 11 February 2015 from http://gallantlab.org/publications/nishimoto-et-al-2011.html.

Gilead, A. 2015. Can brain imaging breach our mental privacy? *Review of Philosophy and Psychology*, 6(2), 275–91.

Golgowski, N. 2014. Scientists use 'mind reading' brain scanners to reconstruct faces people are thinking of. *New York Daily News*, 28 March 2014.

Greely, H. 2004. Prediction, litigation, privacy, and property: Some legal and social implications of advances in neuroscience. In Garland, B., & Frankel, M. (eds.), *Neuroscience and the law: Brain, mind, and the scales of justice*, pp. 115–55. New York: Dana Press.

Greene, J. 2014. The cognitive neuroscience of moral judgment and decision making. In Gazzaniga, M. (ed.), *Cognitive neuroscience V*, pp. 1013–23. Cambridge, MA: MIT Press.

Haynes, J. D. 2011. Brain reading: Decoding mental states from brain activity in humans. In Illes, J., & Sahakian, B. (eds.), *The Oxford handbook of neuroethics*, pp. 3–13. New York: Oxford University Press.

Haynes, J. D. 2012. Brain reading. In Richmond, S., Rees, G., Edwards, S. (eds.), *I know what you are thinking: Brain imaging and mental privacy*. Oxford: Oxford University Press.

Haynes, J. D., & Rees, G. 2006. Decoding mental states from brain activity in humans. *Nature Reviews Neuroscience*, 7(7): 523–34.

Illes, J., & Racine, E. 2005. Imaging or imagining? A neuroethics challenge informed by genetics. *American Journal of Bioethics*, 5(2), 5–18.

Kahane, G. 2008. The art of medicine: Brain imaging and the inner life. *The Lancet*, 371, 1572–3.

Kulynych, J. 2002. Legal and ethical issues in neuroimaging research: Human subjects' protection, medical privacy, and the public communication of research results. *Brain and Cognition*, 50, 345–57.

Langleben, D., Dattilio, F. M., & Guthei, T. G. 2006. True lies: Delusions and lie-detection technology. Accessed 10 February 2016 from http://repository.upenn.edu/neuroethics_pubs/15.

Lever, A. 2012. Neuroscience v. privacy? A democratic perspective. In Richmond, S., Rees, G., & Edwards, S. (eds), *I know what you are thinking: Brain imaging and mental privacy*. Oxford: Oxford University Press.

Logothetis, N. 2008. What we can and we cannot do with fMRI. *Nature*, 453, 869–78.

Miller, G. 2014. Scientists can't read your mind with brain scans (yet). *Wired* 4/29/14. Accessed 10 February 2015 from www.wired.com/2014/04/brain-scan-mind-reading/.

Nissenbaum, H. 2004. Privacy as contextual integrity. *Washington Law Review*, 70(1), 119–57.

Nissenbaum, H. 2010. *Privacy in context: Technology, policy, and the integrity of social life*. Stanford, CA: Stanford University Press.

Parens, E., & Johnston, J. 2014. Neuroimaging: Beginning to appreciate its complexities. *Hastings Center Report*, 45(2), S2–7.

Peterson, A., Naci, L., Weijer, C., Cruse, D., Fernández-Espejo, D., Graham, M., & Owen, A. M. 2013. Assessing decision-making capacity in the behaviorally non-responsive patient with residual covert awareness. *AJOB Neuroscience*, 4(4), 3–14.

Poldrack, R. A. 2006. Can cognitive processes be inferred from neuroimaging data? *Trends in Cognitive Sciences*, 10(2), 59–63.

Rachels, J. 1975. Why privacy is important. *Philosophy & Public Affairs*, 4(4), 323–33.

Racine, E., Bar-Llan, O., & Illes, J. 2005. fMRI in the public eye. *Nature Reviews Neuroscience*, 6, 159–64.

Racine, E., Bell, E., & Illes, J. 2010. Can we read minds? Ethical challenges and responsibilities in the use of neuroimaging research. In Giordano, J. J., Gordijn, B. (eds.), *Scientific and philosophical perspectives in neuroethics*, pp. 245–70. New York: Cambridge University Press.

Racine, E., & Illes, J. 2007. Emerging ethical challenges in advanced neuroimaging research: Review, recommendations and research agenda. *Journal of Empirical Research on Human Research Ethics*, 2(2), 1–10.

Richmond, B. 2012. Main reading and the transparency scenario. In Richmond, S., Rees, G., & Edwards, S. (eds.), *I know what you are thinking: Brain imaging and mental privacy*. Oxford: Oxford University Press.

Roskies, A. 2007. Are neuroimages like photographs of the brain? *Philosophy of Science*, 74, 860–72.

Schleim, S., & Roisier, J. P. 2009. fMRI in translation: The challenges facing real-world applications. *Frontiers in Human Neuroscience*, 3, 1–7.

Shen, F. X. 2013. Neuroscience, mental privacy, and the law. *Harvard Journal of Law and Public Policy*, 36(2), 653–713.

Stoller, S., & Wolpe, P. R. 2007. Emerging neurotechnologies for lie detection and the fifth amendment. *American Journal of Law and Medicine*, 33, 359–75.

Tovino, S. 2005. Functional neuroimaging and the law: Trends and directions for future scholarship. *AJOB*, 7(9), 44–56.

Vanmeter, J. 2010. Neuroimaging: Thinking in pictures. In Giordano, J. J., & Gordijn, B. (eds.), *Scientific and Philosophical Perspectives in Neuroethics*, pp. 230–43. New York: Cambridge University Press.

Wolpe, P. R., Foster, K., & Langleben, D. 2005. Emerging neurotechnologies for lie-detection: Promises and perils. *The American Journal of Bioethics*, 5(2), 39–49.

Conclusion

Michele Farisco and Kathinka Evers

In the Introduction we outlined that infants and DOC patients (which we labeled as the sunrise and the sunset of consciousness) are two complementary states that are particularly useful in order to empirically and conceptually investigate consciousness as a feature of the dynamic brain. Several important points emerge from this book as provisional results that could be fruitfully assumed as starting points for continued work on these challenging issues.

The first part of the book specifically focuses on the intriguing issue of infant consciousness, outlining the impressive progresses as well as the big challenges in exploring the inner world of infants and in implementing a sort of communication with them.

What is it like to be a baby? Hugo Lagercrantz and Nelly Padilla reviewed the fascinating links between brain and consciousness development, particularly stressing that the emergence of consciousness is related to the neurobiological and psychological development of the brain. They outline that infants retain both wakefulness and awareness: there is a biophysical indication of consciousness by the observation of a limited default mode network.

The connection between intellectual and neurological levels is further investigated by Andreas Demetriou, George Spanoudis, and Michael Shayer, who draft a neuro-cognitive developmental theory of intelligence, still to be developed, based on the crucial role of successive expansions of neuronal networks such that earlier networks are integrated into the hub architecture of the networks constructed later.

The intricate cognitive world of infants is increasingly described by contemporary neuroscience: in their chapter, Mohinish Shukla and Vivian Ciaramitaro review the great amount of data arising from behavioral and neurophysiological studies which show similarities and differences between infants and adults in the processing of complex stimuli. Thus we can legitimately assume that there is a cognitive continuity between infants and adults, and that infants possess sophisticated cognitive capacities in several domains. According to the two researchers, it is reasonable to infer from this continuity that the cognitive world of infants has the same basic components of adults: this is an important cue in our exploration of infants' consciousness.

To clear the specific correlation between consciousness and brain, defining principles of interaction between them is essential in order to increase our understanding of infants' inner world, but it has also potential clinical and social impacts, e.g., to predict possible neuropsychiatric diseases early in life and eventually to identify infants at risk and provide targeted intervention. Moreover, notwithstanding the gradual emergence of consciousness as a consequence of brain development, neuroscience has shown that infants retain awareness of themselves and external environment, and their brains are able to consciously process and express emotions. This point, stressed by all the chapters of the first part, has two important implications: infants deserve adequate clinical treatment aimed at minimizing discomfort (e.g., pain); and the infants' "emotional" consciousness could offer a suitable starting point for developing a form of brain communication with them based on internal conscious states involving sensory affective intentions and motor control.

A possible strategy for communicating with preverbal infants is suggested by Karl Sallin, who starts from the hypothesis that conscious states in humans emanate from evolutionary conserved structures, so that it is theoretically possible to compare the human emotional states and behavior with the neural underpinnings of behavior present across species and through this triangulation approach infer the neurobiology of early life experiences, such as pain and discomfort.

The second case-study of this collective book is the condition of patients with DOCs, analyzed in the second part. The review by Carlo Cavaliere, Carol Di Perri, Steven Laureys, and Andrea Soddu stresses the difficulty to behaviorally assess consciousness in these patients and the potential added value of ancillary methods like neuroimaging investigation. The main advantage of this kind of investigation is overcoming the patient's behavioral responsiveness. The potential benefit of the clinical use of neuroimaging is huge: as the authors outline, it is necessary to make this technique an available tool for radiologists and neurologists in the routine clinical assessment.

A potential candidate for technologically assessing consciousness beyond responsiveness is brain-computer interface (BCI). As outlined by Damien Lesenfants, Camille Chatelle, Jad Saab, Steven Laureys, and Quentin Noirhomme in their chapter, to date the significant advances in neural decoding of healthy user-intent as well as the great advances in assessing degree of consciousness in patients with DOCs allow us to use BCI for implementing rudimental but effective forms of speechless communication with positive impact on diagnosis, prognosis, and rehabilitation strategies.

Indeed, the conceptual and technical issues emerging from this prospective BCI mediated communication are extremely challenging, as illustrated by Georg Northoff in his chapter. Empirically, stimulus-induced activity is not sufficient for inferring consciousness: it is necessary the stimulus properly interacts with the resting state activity in order to have consciousness. Conceptually, the inference of phenomenal consciousness from neuronal activity should be distinguished from the inference of cognition from neuronal activity: this is a point often neglected in the discussion about neuroimaging-based communication with DOC patients.

As outlined in the last part of the book, it is important to clarify the ethical and social issues emerging from this prospective cerebral communication. Carlo Petrini stresses the ambivalence of the involvement of infants in research protocols, which is ethically challenging but necessary. The ethically most relevant points are to balance potential risk and benefit and to assure an adequate assessment of informed consent. Neurotechnology can be a good help for analyzing these issues, but more scientific and ethical research is needed.

As outlined by Ralf Jox in his chapter, similar problems arise from the use of BCI for obtaining informed consent from patients with DOCs. Given the strict requirements for valid informed consent in clinical environment, obtaining informed consent from DOC patients appears extremely challenging if not unfeasible, at least to date. The author suggests that informed assent is a possible alternative way for using neurotechnology in order to promote the direct involvement of DOC patients in clinical decisions concerning their condition.

Another issue raised by neuroimaging technology and its prospective applications, e.g., cerebral communication, is privacy. The chapter by Arleen Salles outlines that besides the classical neuroethical approach to this issue (i.e., methodological and metaphysical) we need a more complex analysis of normative issues in order to be able to reach conclusions regarding whether or how brain imaging may constitute a threat to privacy. In other words, we need a new approach to neurotechnology that investigates what is behind our judgment of right and wrong, good and bad.

The main point emerging from this book is that the increasing knowledge about the dynamic nature of the brain can be a starting point for an adequate management of the brain (e.g., in clinical and/or educational contexts), for predicting diseases (e.g., neuropsychiatric diseases) and implement preventive interventions at early stages and for developing new neurotechnology. Within the new neurotechnological tools, the prospective implementation of a cerebral communication is particularly important and potentially raising a positive social impact. The complexity of consciousness as revealed by neuroscience, which at the same time revealed the impossibility to reduce conscious mind to cognition and the important role played by emotion, raises the need to develop new, alternative tools for communicating with speechless subjects. To date, all the tools used are language- and cognition-dependent. In this way they risk to not acknowledge the conscious, even though not cognitive, mind of subjects like infants or patients with DOCs.

Moreover, the inference at the basis of present communication protocols with speechless subjects (i.e., from neuronal activity to mental and conscious activity) is problematic and deserves more attention, both empirically and conceptually:

a) Empirically, the important role played by the resting state in regard to consciousness, as recently revealed by neuroscience, raises the necessity to include the intrinsic activity of the brain and its changes in evaluating the relevance of the brain's reaction to external stimuli as evidence of consciousness: the sole reaction is not enough to infer conscious reaction, because we need a nonlinear resting state change.

b) Conceptually, the gap between brain and consciousness, even if reduced in the light of recent neuroscientific achievements, needs more theoretical work.

Notwithstanding the aforementioned problems, new neurotechnological tools (e.g., brain-computer interfaces) for investigating consciousness and implementing new forms of communication potentially promise huge improvements of speechless subjects' life condition. Among other possible implications, this raises the need to translate the new technologies from the lab to the bedside, i.e., to increase the clinical translation of new neurotechnologies, putting clinicians in the condition to rightly interpret and use these technologies.

Index

amyotrophic lateral sclerosis (ALS) patients
87, 89–91, 136; *see also* locked-in
syndrome
animals 62; decorticated 57, 59
anisotropy: reduction of 74
aphasia 69, 96
arousal: in infants 10, 12; in DOC patients
51, 74
attention 27, 29, 30–1, 94
auditory: in infants 14, 30, 42; in DOC
patients 72–3, 75–7, 87, 90, 92–4, 96,
98, 104, 106–9, 135
autonomy: principle of 121
awareness: in infants 11–3, 17, 22–3,
25, 28, 30–1, 36, 53–4, 56–9, 60,
157–8; in DOC patients 69, 73,
133, 135–6

beneficence: principle of 121
blood oxygen level dependent (BOLD)
signal 9, 72, 146
brain: damages of 70; default mode
network 11, 73; development of 25–34;
reading 143–153; spontaneous activity
of 10–11, 72–3, 110
brain-computer interfaces: assessment of
consciousness via 85–99, 133, 158;
EEG-based 92–4; fMRI-based 92;
informed consent via 136–140

clinical trials 119–120; risk/benefit
ratio of 127–130; with minors
119–132
communication: with infants 54–6;
with DOC patients 89–94
consciousness: assessment of 8–9, 135–6;
attributing 52–3, 106, 110–15;
definition of 7–8, 104; disorders of
69–70; neural correlates of 10, 104–10;
phenomenal 104, 106, 110–14;
primary 54

diffusion tensor imaging (DTI): in DOC
patients 74

electroencephalogram: in DOC patients
76–7, 86–8, 92–4; in infants 42–6

functional magnetic resonance
imaging (fMRI): in DOC
patients 70, 72–3, 86–8, 135–6,
106–7; in infants 7, 8–9, 16,
42–6, 59
functional near-infrared spectroscopy: in
DOC patients 86–8; in infants 9, 13,
14, 16, 42–46, 59

informed consent 133–134; for DOC
patients 136–8; for minors 126–7

justice: principle of 122

language: in DOC patients 73, 96,
159; in infants 15, 44–6, 53, 54,
62, 159
locked-in syndrome 69, 85, 136;
functional 112

magnetic resonance spectroscopy: in DOC
patients 74–5
memory: in DOC patients 92, 95,
104, 138; in infants 14, 15, 22–4,
29–34, 41–2
mind: development of 21–5, 40–7
mind-brain relationship: in DOC patients
104–15, 138; in infants 21–36, 52–3

minimally conscious state 69, 92, 104, 135
mismatch negativity 107–8

neural infantese 56–9
neuroimaging: functional 72–7;
    structural 70–1;
non-maleficence: principle of 121

object representation and individuation:
    in infants 41–3

pain: in DOC patients 75, 98, 135–6, 138,
    158; in infants 13, 59–61, 158
positron emission tomography: in DOC
    patients 75–6
privacy 143–153

smell: in infants 13
sociality: in infants 16, 43–4

transcranial magnetic stimulation: in DOC
    patients 77

unresponsive wakefulness syndrome 69,
    92, 104, 106–15, 135

vegetative state; *see* unresponsive
    wakefulness syndrome
vision: in DOC patients 72, 73, 87, 90,
    93, 96; in infants 14

wakefulness: in DOC patients 106,
    111–12; in infants 12–3